ENVIRONMENTAL REMEDIATION TECHNOLOGIES, REGULATIONS AND SAFETY

SUSTAINABLE DEVELOPMENT

THE CONTEXT OF USE OF INDIGENOUS PLANTS FOR LOCAL ECONOMIC GROWTH

ENVIRONMENTAL REMEDIATION TECHNOLOGIES, REGULATIONS AND SAFETY

Additional books in this series can be found on Nova's website under the Series tab.

Additional e-books in this series can be found on Nova's website under the eBooks tab.

ENVIRONMENTAL REMEDIATION TECHNOLOGIES, REGULATIONS AND SAFETY

SUSTAINABLE DEVELOPMENT

THE CONTEXT OF USE OF INDIGENOUS PLANTS FOR LOCAL ECONOMIC GROWTH

MIROSLAWA CZERNY
UNIVERSITY OF WARSAW
AND
HILDEGARDO CÓRDOVA AGUILAR
PONTIFICAL CATHOLIC UNIVERSITY OF PERU

Library of Congress Cataloging-in-Publication Data

ISBN: 978-1-53612-171-1

Published by Nova Science Publishers, Inc. † New York

CONTENTS

PREFACE

Sustainability has become, in recent years, the keyword essential to any approach regarding development. However, the word itself has different meaning in economic development, politics, cultural evolution and heritage, resource management and other contexts. The main meaning for the purpose of this book is that sustainability seems to be an uncomplete concept since it does not take into consideration the achievements in science and technology which alter our understanding of criteria, like durability, persistence and sustainability itself. In discussions concerning the proper management of resources and changes in the structure of the "basket" of resources that still remain at the disposal of humankind, the key players are human beings, with their consumption models, habits and dependence on products originating from the surrounding environment. It is quite well-known that "sustainable development" is regarded as a concept by which to overcome or at least ease glaring economic and social disparities between the industrialised and the developing countries not only for today, but also for the future. The assumption here holds that poverty and constant shortages facing a majority within the society of each developing country favour an over-exploitation and chaotic use of resources. The over-exploitation showing little or no care for the natural environment is pursued by individuals, small-scale entrepreneurs, and by large concerns that destroy the environment with impunity. We refer here to sustainability as a concept applied to integral rural development in peripheral regions such as the sierra of Piura in northwestern Peru and Viques in the Mantaro Valley of Central Peru.

The development of rural areas in the Peruvian Andes is mainly linked with agricultural activity. Since pre-Columbian times, the management of land for farming has represented the greatest challenge for those inhabiting the region in question. The scarce resources that arable land inevitably represented here was subject to special protection.

In the case of Frías, one of the key indexes where peripherally and marginalisation are concerned is the one relating to the low level of education attained. In the case of Viques, this peripherability is mainly due to its altitude and weather conditions, where crops are limited to only a few.

On the other hand our argument here is that the sustainability is a concept that needs to extend to the management of natural renewable resources, which grow wild in the Andes Mountains and have little attention by local householders. Here, we offer a number of plant fruits which may promote rural development of peripheral rural populations by providing complementary nutritional intakes as well as some extra money to satisfy other human needs. We also refer to a number of medicinal plants widely used by both the rural and urban populations, especially after the growing interest in natural medicine.

INTRODUCTION

While it is true to say that the sustainable development concept grew in popularity in the 1970s, and particularly with the 1992 "Earth Summit" (World Conference on the Environment and Development) in Rio de Janeiro, it was in fact known of still earlier, since its roots can be traced back to the end of the 19th century, when foresters began to take note of the irreversible nature of certain ecological processes arising out of over-exploitative forestry management (Fritz, Huber and Levi 1995). In turn, in 1922, the German urban planner Cornelius Gurlitt launched a debate on "the modern development of the city", proposing that the implementation of programmes for the development of construction in a city should take account, not only of purely technical matters, but also of social and cultural aspects, and these not only from a historical point of view, but also by reference to ongoing transformation processes (Gurlitt, as cited by Petzold 1997: 19). The postulates put forward by Gurlitt constitute an inseparable element of today's definition of sustainable development (Hauff 1987).

From the 1980s onwards, sustainable development began to be regarded as one of the main development paradigms, as well as a fundamental component of regional policy. An exponent pointing to the need for – and means of – introducing the principles of sustainable development was in turn a famous document published by the UN in 1987 entitled *Our Common Future*. The report, from the World Commission on the Environment and Development headed by Gro Harlem Brundtland, accepted that the Earth's resources were running out, to the extent that the opportunities for future development of the planet and humankind on it would depend on wise and rational utilisation and management of the said resources. A further attendant assumption was that the environment should be managed in such a way as to curtail any further dramatic changes in it, and any further impoverishment of it. A pointer given here concerned the role of humankind, and necessary changes in ways of thinking about day-to-day (ad hoc) interests and investments in the economic sphere.

A challenge for development policy thus was and remains the means of introducing the sustainability concept in peripheral regions, in which the key problems to be resolved concerning nothing less fundamental than

The climate change process is already evident all over the world and its effects are perceived by the urban and rural population, especially at poor mountain areas of the Third World. To this event we only rest to design adaptation strategies by optimizing the use of natural resources with appropriate technologies at each ecosystem in particular.

The central Andes are considered as high risk zones vulnerable to the negative impacts of climate change: loss of glaciers that feed the watersheds, which are the water supply that maintain life at lower sectors of valleys; the drought increase due to the greater soil exposition to sun radiation because of deforestation and use of fire to clear the agriculture fields before planting crops; erosion increases due to the absence of forest cover and the intensity of annual rainfall; dryness increase due to higher evaporation rates of soil moisture; etcetera.

Among the adaptation strategies to these changes is a good natural resources management giving attention to the ecosystems of places as support to population that are undergoing a constant increase.

We have already surpassed the first decade of the XXIth Century and the expectations given in the Declaration of the Millenium Development, signed on September 2000 in New York by 189 countries, where they proposed to seek the fulfillment of eight millennium goals up to 2015, have been short in results. A report of the UN in 2013 regarding the poverty solution was optimist (UN 2013). In effect, this report showed that there had been significant improvements in many of the objectives, including the reduction of extreme poverty to half of its existence as it was in 2000. It also reported that there were significant improvements in the provision of potable water, habitability, health attention, education, and others. This report should call us to be happy, but when we make a mapping of the spatial distributions, it appears that the urban centers are the ones that have benefited most and the rurals, especially the ones of the mountain areas are the least favored.

To this socioeconomic condition we may associate the environmental attributes which to many researchers are an inherent part of poverty. Then, the poor rural population needs more and more land to plant their subsistence crops and this goes against the proposed statements to preserve the ecosystems of the headwaters as natural reserves to the production of water. The already mentioned report of the UN (2013:6) alerts us the weak health of the environmental sustainability, because the CO2 production is accelerating, the deforestation continues and by the way many vegetable species, birds and mammals are in the road to extinction.

This scenery invites us to look more carefully to the mountain areas as zones that need to be studied and understood as strategic places to the surviving of societies which shape the identities of many countries of the world. Mountains provide water for use of the population living at lower altitudinal levels, are the ecologic niches of genetic biodiversity, especially plants for food and health, regulate temperature and local atmospheric circulation which allow the existence of diverse ecosystems and an attractive landscape that invites to reflect on it, giving an emotional relax to the urban dwellers.

The underlying topic on mountains, such as in Peru, is rural poverty. Escobal and Valdivia (2004) point out that even when there is a sensible global economic growth, this has not been significant to the rural population, especially in the Andes. The political initiatives to change this condition do not take into consideration the academic advise and continue to apply all fashioned strategies which have been tested as useless to the Andean rural (ibid; Bragg Egg and Mendiola Vargas 2000:395). This poverty is strongest among the ones of Andean heritage who live in remote rural places; and it is calculated that 20% of the highlanders of Peru are in this condition of extreme poverty (FIDA 2013).

Poverty in the Peruvian sierra correlates positively with chronic food insecurity of the rural, where many households live on subsistence basic food dependent on the seasonal summer rainfall. To this, we may add the fact that there is a kin relationship to traditional crops and show resistance to any change even when they have other native crops at hand reach.

The poor appreciation of the local environmental offer has, to some extent, the support of the education system, where the teachers are pressed to fulfill the academic program given by the Ministry of Education with books elaborated in Lima, to be applied at urban environments which know nothing of the more rural local ecosystems nor to their diversity.

LIST OF ABBREVIATIONS

AWWA	American Water Works Association
CCL	Contaminant Candidate List
EDF	Environmental Defense Fund
EPA	Environmental Protection Agency
GAO	General Accounting Office/Government Accountability Office
GAC	Granular Activated Carbon
HRL	Health Reference Level
MCL	Maximum Contaminant Level
MDEQ	Michigan Department of Environmental Quality
MHS	Marine Hospital Service
NAS	National Academy of Sciences
NDWAC	National Drinking Water Advisory Council
NPDWR	National Primary Drinking Water Regulation
NRC	National Research Council
NRDC	Natural Resources Defense Council
PCCL	Preliminary Contaminant Candidate List
PHS	Public Health Service
SAB	Science Advisory Board
SDWA	Safe Drinking Water Act

PART 1. SUSTAINABLE DEVELOPMENT IN THE PRACTICE

Chapter 1

THE TERM "SUSTAINABLE DEVELOPMENT" AND THE CONTROVERSIES SURROUNDING IT

Notwithstanding the elaboration and publication over the last half-century of countless official publications invoking a need for sustainable development principles to be put into effect, a host of scientists, politicians and planners continue a (heated) debate on the significance of contemporary society's understanding and heeding of the principles in question (Petzold 1997). An idea proposed several decades ago has become one of the most important paradigms in development policy, and a key to analyses of directions to development on different spatial reference levels. Indeed, the theoretical and planning-related discussions on the subject are participated in by representatives of different scientific disciplines, notably geographers (Gutry-Korycka 2005). It might therefore seem that the identity of the issue under discussion is well-known and understood, which is to say that we now know what sustainable development entails.

It was with such a conviction as to the existence of some broad knowledge of sustainable development among the inhabitants of today's world that a decade of education in the name of it was launched by UNESCO, with this period in fact coming to an end officially in 2014. This would logically imply that the last ten years have already seen a broadening of society's knowledge on the subject, its objectives and the effects of its implementation. Indeed, all age groups should have found themselves brought within an educational process allowing us all to face up to and address the challenges associated with sustainable development that have been identified at international fora.

The sustainable development idea in fact assumes that the socioeconomic development ongoing in the contemporary world will proceed in such a way that key existing features of our natural (but also our social and cultural) environment, and hence our "surroundings" in the broader sense, will remain in such a state of preservation that the generations coming after us will be in a position to use and draw benefit from them just as we have. While this relatively-well known assumption seems clear and obvious, it does not of course make full reference to achievements in science and technology which may as time goes by alter our understanding of criteria like durability, persistence and sustainability itself, when it comes to various different elements of the environment – up to and including the conditions underpinning the development of agriculture.

However, the real truth is that now, as from the outset, the discussion on sustainable development has entailed many and various conceptualisations and ways of understanding the very concept, let alone its manifold different aspects, which are treated differently at both the interpretation stage and when actual procedures or activities are put into effect. These differences are such as to ensure that quite disparate directions can be followed as efforts to achieve sustainable development move forward.

One clear and readily noticeable evolution of the concept has proceeded from approaches entirely (or almost entirely) focused on the natural environment and the need to protect or secure its air, waters, soils and natural resources (Czerny 2005) in the direction of approaches that now seek to integrate many different tiers of life and types of human activity, in particular assigning value to (and hence encouraging the protection of) elements of our heritage, be this natural, cultural or even political.

Another division to the approaches to sustainable development entails analysis of local and regional potential where the stimulation or continuation of growth is concerned. What it therefore ushers in is a critical analysis seeking such means of proceeding, and of applying techniques and technologies, as will cause least modification to natural environmental conditions and (in theory at least) to existing social and cultural conditioning also.

We thus arrive at issues of the dynamics to socioeconomic change, proceeding in so doing from an *a priori* assumption that a given fragment of territory is experiencing such changes constantly, with the result that geographical space is also undergoing change. A very simple diagram can help show the relationships pertaining in a territorial system which is defined in this kind of way.

In this understanding the territory and its natural conditions are given, and they most often change over long periods of time. However, as will be noted later, the territory involved in given considerations can also be treated as a variable where development is concerned. In any case, in the simple model of development, the two most important factors determining the trajectory, dynamics and structure to favourable changes in a given region or territory are seen to be human capital on the one hand and economic capital on the other.

A more complex model of development will bring in a series of further variables, including those of importance to sustainability like historical conditioning (the tradition of a historic region), psychological and emotional features (not least prejudices and schemes where ways of thinking are concerned), and cultural conditioning (within which religious conditioning can be of key importance in determining, setting or shaping models of living and types of conduct).

Is a return to a simple development model still possible, as consideration is given to the structure of sustainable development and its main actors? In general, yes, though with the proviso that the actions of both human capital and economic capital should be in line with sustainable development principles. So what should this entail? Answer:

1. an awareness (among all members of a given society) that any kind of human action (or intervention) whatever in geographical space causes change to – and *in extremis* the destruction of – the existing natural environment.
2. the use in all kinds of human activity of those techniques and technologies that do least to change the character of the given region, this extending to its environment, obviously, but also the skills of its inhabitants, customs and habits, and so on.
3. financial (economic) capital invested in a given region that first and foremost takes care of the interests, wellbeing and living environment of the inhabitants of the said (or any) region.

In the face of the above, can the sustainable development concept be adhered to as we seek to develop peripheral regions (in which by definition raised income levels and an improved quality of life are called for)?

From the last decade of the 20th century onwards, against the background of the ever-more widespread discussion of worsening environmental, economic, political and social problems, and in the face of crises and the search for ways of exiting from them, the subject of sustainable development came to be seen as the most frequently-contested reflection on the future of development. For this reason also, for almost three decades now, both the terms "sustainability" and "sustainable development" and the concepts considered to underpin them have been among those most popular and most often resorted to in public discourse. Since 1992, when they came to the fore as matters of key importance arising out of the 1992 Rio "Earth Summit" (UN Conference on the Environment and Development), they have scarcely been off the front pages of newspapers, and have also been a constant presence in scientific journals, as well as being the subject of political debate.

However, from the outset, the introduction into academic deliberations of the sustainability concept and term alike was associated with a great many doubts – and a great deal of criticism – even to the extent that the potential for them ever to gain real-life application seemed to be under question.

Since the 1987 "Brundtland Report" itself did not elect to put forward any more precise definition (Schneider 1993), those wishing to develop an understanding of sustainability first elected to make reference to ecology, in the hope that that domain would prove able to shed fuller light on the matters involved.

In the meantime the concept (or more precisely the very term) "sustainable development" started to be treated as a source of hope for a better future that was proving fascinating to researchers and politicians/policymakers alike. By that point, civilisational development had already come to be identified with the need to maintain such natural conditions on Earth as would be able to sustain the lives of the human population; this in turn encouraging a search for consensus over such matters as the identification of common goals and the possibility of introducing instrumentation to allow for the achievement of these objectives.

Nevertheless, a discussion on sustainable development pursued for almost half a century now has thrown into sharp relief the weakness of the scientific argumentation in favour of this kind of development, not least the difficulty experienced with resolving many of the key environmental problems, such as those involving the capacity of different ecosystems to "take what is thrown at them" (in terms of various kinds of harm done and physical matter – e.g., wastes – introduced) and still rebound. It has been made clear in practice that issues of this kind are not merely of importance or some relevance to politicians, but are rather of leading importance more generally.

Polish policy documents and scientific studies alike make use of the term *rozwój zrównoważony* as an equivalent (translation) for "sustainable development". Yet from the moment this Polish version made its appearance, there have been two conflicting views when it comes to the naming of a process whose assumptions and means of achievement need to be written into both academic studies and planning documents. While many are for the above term (which nevertheless makes some at least indirect reference to the idea of "balance", others – perhaps drawing on the French-language version for the term – advocate use of *rozwój trwały* (where the word *trwały* in many contexts alludes to permanence, durability, persistence, and therefore also to something being (or capable of being) maintained or sustained or continued with in the long(er) term). Ultimately, it would seem that it is more typical for *rozwój zrównoważony* to be used, and so it has been this term employed in the Polish version of this text. As Haber (1995) stated: "Those who read the English-language environmental literature between 1980 and 1990 met with descriptions of relations important to the functioning of the natural environmental system, which were characterised by a certain permanence [continuity and constancy]. However, occasionally it was possible to meet with the terms *sustainable*, *sustainability*, *sustainable use* or *sustained yield*, though the application of these was confined to a new ecological current in the natural sciences that was not followed at all outside a narrow group of specialists. To such a degree was this issue omitted from scientific debate that no definition was included in the oft-cited Carpenter ecological dictionary published in 1962" (Haber 1995:17). This changed fundamentally in the 1980s after the Commission on the Environment and Development under Norwegian politician G.H. Brundtland made its April 1987 presentation of the Report entitled *Our Common Future* (WCED 1987). In it, "sustainable development" was adopted as the main line for future environment policy worldwide.

So how did the choice of the term – and its huge popularity in both scientific and political circles – come about?

At the end of the 1960s, quite unexpectedly and suddenly, environmental protection issues became transformed into some of the most important subjects of political debate in industrialised countries. The USA's National Environmental Protection Act with far-reaching provisions concerning the protection of the environment came into force in 1969. 1970 was in turn declared *European Year of Nature Conservation*, with the aim being to raise the general level of awareness of the Old Continent's inhabitants, as regards the scale of the destruction that nature was experiencing, and the consequent need for Europe's inhabitants to mobilise in the name of the environment. Many countries established Ministries of the Environment. 1972 then saw the United Nations organise its first environmental conference – in Stockholm, this in fact being named the UN Conference on the Human Environment. It was in the course of that that serious controversies arose between the developed and developing countries, with the latter accusing the former of seeking to introduce principles that would make further development difficult and therefore help keep in place a relationship founded upon dependence.

The negotiations taking place following the Stockholm Conference showed that new problems associated with the environment could not be resolved if account were not taken of the social and economic requirements (or demands) of those inhabiting a given area or region, and most especially inhabitants of what was then still known typically as "The Third World". In the face of the new challenges posed to the international community, and with a view to developmental disparities between the First and Third Worlds being reduced, a special UN Commission was called into being with the task of developing assumptions that would underpin "environmentally sound strategies". It was to make it easier for different economic and political actors to devise and implement mechanisms that would support "just development". An almost immediate response to that came with the appearance and widespread use of the term "ecodevelopment" (Strong 1980). However, this was soon abandoned by most international organisations, on the basis of the fact that "eco" might equally well be seen as an abbreviation for "economic", as opposed to "ecological", which would then have given rise to misunderstandings as the issues were discussed. Nonetheless, it did not prove possible to eliminate use of the term entirely, and it is paradoxically now the case that more and more international documents and academic discussions are making use of "ecodevelopment", given the way that the latter is felt to extend more widely than "sustainable development".

However, at the end of the 20th century, there was no doubt whatsoever regarding the popularity of the term "sustainable development". Indeed, it would continue to dominate the debate for some three decades (Sachs 1992). It may be that the popularity of the term was contributed to by a sentence in *Limits to Growth* in which the author writes: "... it is possible to alter these growth trends and to establish a condition of ecological and economic stability that is sustainable into the future" (Meadows 1972, cited by Heinen 1994). It is interesting that the term "sustainable development" did not gain its first application thanks to the World Commission on the Environment and Development, but rather as early as in 1980 – in the IUCN's *World Conservation Strategy*.

It was nevertheless the Brundtland Report which encouraged the organising of a second UN international conference on the environment, i.e., the United Nations Conference on Environment and Development (UNCED) or "Earth Summit", convened in June 1992 at Rio de Janeiro. The participating countries there undertook to devise a broad plan of action for sustainable development (called *Agenda 21*), with there being 40 points and recommendations for the UN, including one that a Commission on Sustainable Development should be called into being. This appeared in 1994, and its first head was the then Minister of the Environment of Germany, Mr Klaus Töpfer.

The Spanish language makes use of the English-influenced *desarrollo sustentable* or *desarrollo sostenible*, while French has *développement **durable***. In the German documents relating to the Brundtland Report, sustainable development was in turn translated as *dauerhafte Entwicklung*, in which the leading idea is one of permanence. In the 1990s *dauerhaft* came to be replaced by *tragfähig* – which has much more the meaning of something that is doable (capable of being implemented). However, use is first and foremost made of a term brought in from forestry, namely *nachhaltig*, wherein the idea is of something that can be continued with at a steady level for longer periods. Since the application of the term in English to agriculture and forestry comes out as "sustained" (as in "sustained yield" – i.e., one that can be maintained for longer periods), a German version more akin to "sustained development" is also sometimes in use.

Chapter 2

ON THE CONCEPT

Since the 1990s the term (and underlying concept) of sustainable development has become the aspect arising most often in discussions regarding the Earth's fate, the management of natural resources, energy policy, the safeguarding of food security and places to live, and so on. However, ideas on the accomplishment of sustainable development assumptions have first and foremost allowed for a development and strengthening of all those ecological movements able to put political pressure on the governments of different countries, with a view to their devising and implementing policy on environmental protection, as well as the instrumentation allowing for effective control over the exploitation of natural resources (Roorda 2012, Elliott 2013 4th edition, Romero Rodríguez 2012). Analyses concerning the structure and dynamics to contemporary use of natural resources for economic purposes which came into existence in time for the Rio "Earth Summit" (i.e., in the early 1990s) made it clear that the prevention of excessive change in the natural environment – and the guaranteeing of its appropriate quality for both present and future generations would be best served if the main focus was on raw materials and their use (Henseling 1995). This rather one-sided way of thinking did in fact evolve over time, to the point where activity in the name of sustainable development has now become multifaceted, involving a wide range of entities, and extending far beyond the field of raw-materials and resources use (Czerny and Córdova Aguilar 2014).

The 1992 "Earth Summit" (World Conference on the Environment and Development) convened in Rio introduced many provisions obliging economic, political and social actors to put the principle of sustainable development into effect. Thus alongside political will (whose generation represented the first step towards the adoption of the aforementioned principle by different countries), there was a crucial need for the instrumentation essential for the process to gain momentum to be first devised, and then introduced. Within this, there were methods concerned with the governance of the sustainable development process. However, as was soon to become clear, the adoption of a comprehensive strategy ushering sustainable development into each of the processes present in each different subsidiary field is an extraordinarily difficult matter. So much so in fact that only in recent years have certain serious solutions begun to take shape. Zerka (2015) cites those adopted in Norway and Germany as examples, writing in the case of the former that the Biodiversity Act drawn up "denotes a series of principles for the sustainable use of natural resources that all Ministries must conform to" (Zerka 2015: 2). Where Germany is concerned, the introduced energy reform or *Energiewende inter alia* includes: "The development of renewable energy sources, a departure from the use of nuclear power, ambitious emissions targets, and the involvement of citizens in the processes by which energy is generated and distributed". (Zerka 2015: 2). Representatives of the German Government claim that the reform is a success, not only for energy policy, but also for the strategy of sustainable development. "The success to date with the reform is inclined to be accounted for through the prism of society's departure from modernisation understood in a "linear" fashion to a "reflective" version that has required the building of a hard-to-achieve consensus between citizens, industry, the political parties and the regions in Germany. And a key role in this process has also been played NGOs and numerous advisory bodies that intermediate between civil society and the authorities." (Zerka 2015: 2).

Thus, as it addresses matters of resource management, the literature on sustainable development (Henseling 1995; Roorda 2012, Elliott 2013 4th edition, Romero Rodríguez 2012) most often accepts that the process encompasses the means of utilising renewable and non-renewable natural resources, as well as the consequences of such activity, in the form of wastes generated and the possibilities for them to be disposed of or utilised. The research topic formulated in this way can then represent a guideline or basis for the devising of sustainable development strategies and policies at the national and regional levels, as well as serving as a starting point from which to develop an overall policy for the management of natural resources (that does not always proceed in line with sustainable development principles).

Under the concept espoused by Ewards (1992), care for the developmental opportunities of future generations should denote natural resources being retained at the same level both quantitatively and qualitatively. This assumption in turn leads the author on to several principles that should be addressed by all those formulating the basic guidelines on managing natural resources (Ewards after Henseling 1995:82). These are:

- that the time needed for renewable resources to be degraded (consumed) should not exceed the time taken by their regeneration;
- that the level of resource use available to humankind should not exceed the possibilities afforded by the natural conditions in the given place, when it comes to absorption (or assimilation);
- that a decline in the availability of non-renewable resources may only be justified where the means of exploitation still guarantees no worsening of the existential conditions of future generations (Ewards after Henseling 1995:82).

On the basis of the principles cited above, many countries have adopted and developed bases for the management and utilisation of raw materials, as well as strategies by which they may be economised on and protected (*ibidem*). Proper resource management – and the achievement of one of the key goals of sustainable development that that denotes – entails monitoring of the rate at which, and scope within which, raw materials are used. This is especially important in the case of renewables, whose renewal safeguards the needs of present and future generations, provided at least that the demand for them does not change markedly. This kind of approach goes along with the so-called efficiency of nature, whose resources available to the economy and humankind are also known as ecological or environmental capital (Czerny and Córdova Aguilar 2014). The efficiency of nature may not be measures solely in terms of its capacity to supply raw materials for production, being also understood to embrace the function of maintaining life on Earth, the structure in this way being particularly sensitive to the disruption of relationships that existed originally. The principles for the management of nature's resources to meet the needs of the economy is underpinned by the assumption that, should these needs be played down and not taken seriously enough, the conditions for management and development in the future (if also for a short period) will worsen, while defined bases for the economy will be disrupted for ever or for a very long period, as a result of excessive burdening of a sensitive environment through over-exploitation.

The existence, recognition and exploitation of natural resources represent key aspects of human-environment relations and a precondition for human life and economic activity. The means of management applied currently – in which huge quantities of resources (materials and raw materials) are consumed, with a high level of trade in them also taking place between countries – ensures the destruction of the natural bases upon which the economy of this kind is actually founded. Rapid ongoing change to the structure of a whole range of different resources that the economy makes use of denotes ever-greater and more serious global consequences that are very much irreversible and nature and ensure that overburdened and unrenewed ecosystems are not in a position to respond or recover. Heavy burdening of the environment (human pressure on resources) is thus linked with growing economic consequences and costs, as augmented by social (and health-related) costs, losses of an aesthetic and cultural nature, and a level of destruction that is incapable of being remedied.

Kopfmüller (1995: 106) stated that sustainable development could be defined as a process by which efforts are made to attain three fundamental objectives, i.e.:

- a worldwide limitation or curb on the overall consumption of materials (resources), energy and land;
- an improvement in the material and non-material conditions characterising the lives of a large part of the inhabitants of the Earth (especially in countries of the global "South");
- a raising (through education) of people's level of awareness that the use of resources limits possibilities for the development of future generations, while not denoting a loss of (resignation from) the quality of life enjoyed hitherto by our planet's inhabitants.

Sustainable development was defined in some detail in both the Brundtland Report and subsequent UN documents issued both before and after the Earth Summit. This is at least how it seemed to scientists and politicians involved in the preparation of the final documents. In the course of discussions subjecting the concept to criticism – and most especially pointing to difficulties with its application in practice – a number of features and criteria were identified as having been assigned a relatively clear scope within the wider framework of the concept, even if this scope was not entirely free of ambiguity. Sustainable development has thus been taken to represent a certain normative concept corresponding with project tasks defined more precisely. As would be entirely typical, implementation of each project task has to be preceded by an analysis of the overall conditioning of development in a given area – in order that concrete technical measures can then be brought into effect (Huber 1995). Thus, while the issue of sustainable development was one which arose out of academic debate, the worlds of both politics and economics have sought to treat it as a project having real-life application. Today, therefore, science has the role of leading discussions where visions of sustainable development and its application differ. Unfortunately, however, there is a lack of consensus in the normative sphere when it comes to the scope and level of detail of data to be used in measuring the degree of sustainability that sustainable development has achieved. Unanimity is also lacking regarding the ways given countries or regions should be developing if they are to meet (or at least meet more fully) sustainability criteria-cum-requirements. It proves far easier to say what is not sustainable development, and world leaders have come to an agreement that the kind of global consumption of resources and energy that we are seeing at present cannot be maintained or continued with. There is of course no such consensus where defining the causes of the excessive use of resources is concerned, let alone with determining preventative methods and measures.

From the mid-1990s onwards, sustainable development was ever-more likely to be treated as a subject for research, not only in the natural sciences, but also from the point of view of the social sciences. Research initially focused on environmental protection and education began then to take in analyses of the aspects whereby development was to be made more sustainable, also in the context of greater balance and enhanced equality. A further, related thrust saw matters both environmental and social looked at from the point of view of the economy (with the resultant field being environmental economics), as well as in regard to the legal and administrative aspects (notably legal measures that would seek to ensure the wise use and management of resources). Further work, developed in the last two decades in particular by authors including Fritz, Huber and Levi (1995); Jacques (2015) and Czerny (2015) seeks to achieve interdisciplinary analysis – and interdisciplinary action – in the name of sustainable development. In this respect, the search is on for those criteria that can define this kind of development properly, with research then pursued to determine the degree to which systems studied comply with them (or more probably in reality, the degree to which the criteria might in future be complied with, or indeed the possibility of their ever being complied with).

In recent years in particular, much work has been devoted to the discussion of indicators of sustainability, i.e., the measuring of the phenomenon and the comparative methods by which it might prove possible to display the whole spectrum achieved by the phenomenon in different regions and different spheres of life (Zottis, Russo and Araujo 2009).

The nature-related definitions of the term sustainable development are seen to differ from one another in regard to the system of reference, e.g., forest, seas and oceans, climate, the Earth's surface, rivers and lakes, plant cover, etc. Studies in this field have been appearing in both the biological and Earth sciences, and their assumption is that these and other ecosystems need protecting if they are to persist and so be available to serve future generations. Detailed analyses of the cycling of matter in studied systems are then to show how much disturbance or distortion of the natural balance existing within them can be sustained or tolerated without far-reaching changes of structure being precipitated, or else total disappearance. For biologists for example, a system is in balance when the activity and intervention taking place within it does not lead more matter to leave the system than enters it in compensation (*ibid.*).

What then does the ecological interpretation of sustainable development's conceptual scope look like? Irrespective of whether it is continuity, stability, persistence or durability that is of greatest importance, how does the aforesaid sustainable development actually look? First off, it was an analysis of natural ecosystems like forests or mountains or whatever, that took place, with the search being on for essentially unchanging structure, notwithstanding ongoing processes of birth and death among the organisms constituting components of the system under study. Even in the wake of natural disasters like fires, storms and floods, the systems in question are seen to be regenerated (returned to equilibrium) rather rapidly, with composition not modified excessively in the post-disaster circumstances as compared with those existing before. Such "stability" of natural ecosystems is the "ecological balance" whose structure and processes have been the subject of research for many decades (Gigon 1984, Schwegler 1985). It is thus analyses of the composition and internal relationships of ecosystems that can offer a basis for action in the name of environmental protection and nature conservation, with the aim of ensuring that natural ecosystems are preserved.

For example, Fritz, Huber and Levi (1995) claim that humid forest in the equatorial zone represents the best illustration of the process described above. It can be regarded as almost a prototype for equilibrium or balance in nature, given that, while oxygen, water and small amounts of mineral salts are in flux here, biomass is steady and being regenerated constantly. A question remaining open concerns the ease with which forest reproduces, the question being as to whether biomass really remains constant or actually rises as the forest grows (Fritz, Huber and Levi 1995).

Sustainable development does not relate solely to a process of more prolonged persistence, being also concerned with the level of resources, and the management thereof, in nature. Resources are required by all of the Earth's organisms if they are to live, but the availability and magnitude of the former are also dependent on cycling in nature. The main resources are nutritional components and energy (with the means of further existence in the case of animals (including humans) being at the same time carriers of energy). Humans are part of the (any) natural ecosystem and their first role of hunter-gatherer resembles rather closely those played by other animals. At the same time, humankind – more uniquely – has spread steadily across the entire globe, with activity coming to be differentiated in this way, as adaptation to differing conditions took place. Human beings were always so very linked with the conditions in the environment in which they lived that they were in a position to adjust to circumstances capable of being considered extreme by planet-Earth standards, in this way surviving periods of cooling and even the Ice Ages. This survival or persistence could itself be taken to denote a kind of sustainability (Fritz, Huber and Levi 1995). Yet the last 10,000 years of that ongoing survival has seen human beings change almost everything else in nature to a huge degree that is often dubbed or regarded as "irreversible". And the evolution associated with human activity on Earth, in concert with the accompanying social and cultural change, has distanced our species from "sustainability" in the balance-related (as opposed to the persistence-related) sense. We are thus far from what some authors would see as the ideal model for the human-environment relationship, which would have people as component elements of given natural environments, within which they develop in manner sustainable for the latter, as well as for themselves (*ibidem.*). To put it mildly, this would not seem to absolve science of the need to analyse "sustainability" from the point of view of the influence on nature of technological and industrial (as well as cultural) change.

Through studies of forest ecosystems, Kurth (1994) sought to determine the extent to which the ecological views on equilibrium and sustainability are of importance or relevance to today's academic discussions, let alone for planning purposes, where actual application of the sustainable development concept is concerned. Kurth wrote that: "The balancing of development is with a term taken from ecology. However, in the broad sense it is the same as management in the defined environment" (Kurth 1994: 37). Now on the basis of a discussion pursued in the forest sciences there could be no doubt that Kurth's balancing of management or of the economy was in essence also a commitment or obligation on the part of society, given that it reflected "a principle that ongoing consumption be eschewed in order to safeguard existence in the future" (Kurth 1994: 37). However, for the practical application of that principle, Kurth states that the forestry literature has had rather little to say about more precisely defined conditions for when we are dealing with forest management in balance (or even sustainable forest management) and what influence this has on rural areas. This is all the more so given how frequently forest management takes place within a monoculture or near-monoculture and by definition cannot therefore be characterised by sustainability (Kurth 1994: 37, 45).

An interesting proposal was advanced by Kopfmüller (1995: 106), who asserted that the scientific and political discussions held up to that point had had little effect where the practical implementation of the sustainable development concept was concerned. In the first place, the cause of the failure needed to be sought in the rather narrow treatment afforded the issue in the case of the developing countries. There the assumption adhered to has been that difficulties with development result from poverty and population increases, irrespective of the broad political, social, cultural and economic context in which these processes have been ongoing all the time. In the second place, many analyses have focused solely on the technical and administrative dimensions to development, with no account being taken of the regional and local specifics. In contrast, in the view of Kopfmüller (1995: 106), the sustainable development process must take in all of the dimensions referred to. In this way, it should "take account of environmental, economic and socio-cultural components, while it must also relate to stages in the process whereby policies are introduced – from the defining of the nature of a problem via the formulation of objectives, choice of means (ways of acting) and implementation, though to the monitoring of the effects of implementation. Finally, reference must be made to all levels of the territorial division from the local through to the global, with account taken of existing interactions between industrialised and developing countries as an effect of policies once pursued and means employed by the former." (Kopfmüller (1995: 106).

Kopfmüller (1995) further states that a policy of this kind should be a long-term one, and should relate to dynamic processes that are temporally and spatially variable. Each time it should make reference to regional and local conditioning of development, and be open to new developmental approaches and strategies that are appearing.

Chapter 3

SUSTAINABLE DEVELOPMENT AND PROCESSES OF ECONOMIC DEVELOPMENT

The sustainable development concept views ecology as closely linked with the economy, with these being the two key components of a system generating the leading axes when it comes to principles for rational management in a given area, with acceptance of and compliance with the rules by which the natural environment functions. This does not denote that the interdependence is a harmonious one, as in general it is possible to observe severe conflicts as regards objectives (profit on the one hand and wise and restrained use of resources on the other). Equally, common objectives can be mentioned (like the rational use of given resources that are only regenerated over defined time intervals). In general, the sustainable development system should give rise to a situation in which – in the long term – relations based on similarity between environmental and economic interests will be ever-closer.

In the report by G.H. Brundtland, economic and environmental objectives are linked simultaneously with the accompanying social goal of a fair distribution of resources (Huber 1995). If such a concept is consented to, then we ought to refer to sustainable and equitable development. Leaving aside the fact that such a term does not sound especially good, the objective of a just division of resources can be seen to move beyond either economics or environmental policy, instead becoming a domain for political decisionmaking first and foremost. The addition of this last element to the definition of sustainable development has the further effect of making the latter entirely unachievable, at least at this time, especially in the countries of the global "South".

Furthermore, a term relating to the ecological aspects of development can be applied in place of the term resources management, since the latter is encompassed by the former. Such an understanding arises out of the traditional approach to sustainable development adopted by foresters and ecologists, for whom the concept was always linked with the proper management of resources (forests, waters and so on). This is all the more the case given that debate on sustainable development is indeed underpinned by a fear of resources essential to development running out, with it therefore becoming necessary to manage appropriately both the resources themselves and the overall economy making use of them, and indeed the overall natural environment.

Sustainable development might thus be treated as a broader conceptualisation of questions concerning development in general that arise at the national and global levels, at which economics takes the laws of ecology into account. Huber (1995) states that anthropogenic and geogenic processes are equally accepted in this connection (Huber 1995: 33).

Apart from addressing fundamental questions regarding the need for environmental protection on the world scale, and supplying guidelines for all the world's governments on the implementation of sustainable development principles, the documents from the Rio Earth Summit say much about principles in line with which national economies – including those of developing countries – might be modernised. However, looking back on it now, it is clear that ideas that seemed progressive and achievable at the time are in fact hard to implement at all, and are almost exclusively being tackled by the governments of developed countries. The huge pressure to achieve economic growth at any cost has ensured that we witness ever greater harm being done to the natural environment, with a reduction in the area of natural ecosystems and huge amounts of pollution in different parts of the world. The governments of countries worldwide striving to achieve higher incomes for their societies are not inclined to accept and bring into force effective environmental protection instrumentation, especially where concessions are being granted to the multinationals to exploit mineral resources (as, for example, with Yanacocha and Corani), or else where large-scale infrastructural developments are being pursued.

The principles for modernisation set out in the Rio documents also address the matters of world trade and development policy. According to their authors, these matters might only be taken account of in a document as general and capable of the superficial reconciling of different interests as the Rio Declaration. The two postulates were and remain a subject for discussion and global agreements. For example, one of the key postulates as regards the sustainable development concept concerns the fight against poverty in countries of the global "South". In these it is assumed that a formulated modernisation concept will make environment-friendly development a possibility. The sustainable development concept can also be said to have offered a first bringing-together of social and environmental issues as a set interlinked into a "policy for the Earth" (*Erdpolitik* – E.U. von Weizsäcker 1989 after Huber 1995: 33). Hence relationships and linkages of both an economic and an ecological nature came to be a global issue, rather than merely a regional or national one, as they had been before the Earth Summit.

Agenda 21 was not just to establish a global environmental protection system – to prevent the destruction of ecosystems extending beyond state boundaries and being of fundamental significance to humankind. Rather it was also to use a philosophy of care for the environment to achieve global development and enhanced wellbeing, with transfers of knowledge and technology, and a global economy and trade system adjusted to serve universal development.

As has been noted already, the widely-applied term "sustainable development" is most often related to natural phenomena, and notably nature conservation and rational use. However, it is characterised by a certain similarity – impossible to ignore – with the physiocratic economic principle of rising internal relations (cycling) under circumstances of balanced external relations (Fritz, Huber and Levi 1995). If the true situation is as the authors in question seek to describe it, then a research topic important from the point of view of geographical (spatial) analysis arises, i.e., involving the territorial and temporal limits of the system.

From the point of view of the natural sciences, sustainable development and the process of growth are not mutually exclusive, or contrary to one another. Just because development occurs, this does not necessarily mean that balance in nature is disturbed. This is perhaps a trivial observation, but the simple production of a system in which anabolic and catabolic processes remain precisely in balance with each other is something that rarely happens (Fritz, Huber and Levi 1995). In a normal system it is not the case that there are periodic and aperiodic (irregular) variations, in which short-term and long-term processes of growth or development overlap with one another in a given system. The best-known examples here are perhaps such phenomena as the annual rhythm to the growth and dieback of vegetation, or the systems of better or worse trading conditions in the economy (Fritz, Huber and Levi 1995).

In the social sciences, a concept for sustainable development appeared in the context of modernisation theory (including quite often in critical analyses thereof). Proceeding on the assumptions inherent to that theory, it is assumed that scientific and technical institutional development is taking place, and within the framework of a process of modernisation understood in this way there is a further process of revaluation and a new approach to matters of resources and the natural environment. Under the theory in question, the environment is mainly treated as a "resource" – i.e., an element of the social and economic system that is joined by other elements in generating a basis for the existence of conditions supporting progress (modernisation), as well as playing an active supporting role in that process. The concept of social development and attitudes to resources (i.e., to the natural environment) is based around the concept of profound social change induced by industrialisation and the reformulation of the attitude to resources.

The process of industrialisation prevalent in economically-developed countries almost to the end of the 20th century was treated as an example of unsustainable development, being based around the assumption that fossil fuels (especially coal resources) would not be running out any time soon in the world as a whole, and were able to constitute the main motor underpinning progress. Attention was then drawn to the idea that such a process of development did not correspond with modern concepts for the protection of the environment and the use of resources, and could not be regarded as sustainable, with the result that people started to postulate the need to follow a path of restructured, reformed and self-modernising development (Roorda 2012).

As a number of publications have made clear, discussion on the concept of sustainable development as such, the ways in which it should be understood, and the possibilities for it to be put into effect, have assumed a similar character to the discussions concerned with growth and development that took place in the 1970s (Roorda 2012, Jacques 2015). Then as now it is important to determine a scope for the definition that extends wide enough to allow different groups in society and units of administration to match their vision for development with the global vision of progress and increased wellbeing combined with a respect for the natural environment all around us. What remains of greatest importance in this discussion (and is at the same time a motor for change taken account of by the natural and social sciences alike) is growth – as measured using precisely-defined economic indicators comparable in all countries around the world. Thus the worldwide discussion of sustainable development has points common to all participants, such as growth measured quantitatively in this way, but also subjectively (qualitatively), by reference to perceived improvements in living conditions or quality of life.

Development – treated as a holistic process made up of many different factors and kinds of conditioning – is seen to take place in discontinuous fashion, with an influence on its rate and variations being exerted by both external and internal factors in the given region (each territorial unit of administration). Relations that are optimal from this point of view create a balance between the natural environment, the needs of society and economic activity. Alongside the sustainable development concept, there has been a defining of areas of action of a political nature whose reach is a global (worldwide) one. There has thus also been a designation of subject matter taken to fall within the relevant political and environmental paradigms. Where planning and political practice is concerned, the process of defining these should also have seen tasks pinpointed for different countries, as regards both politics and the economy.

A process of definition has also encompassed the subject matter of scientific research into sustainable development, with matters of importance to be put into effect within the framework of regional policies also identified, including in connection with defined technical and organisational conditions, as well as economic ones.

Production strategies in agriculture – especially traditional agriculture – are defined and adopted by farming communities over time. Decisions in this regard are based around internal and external factors that farmers and livestock breeders take account of as they plan their activities in a given season and for the upcoming period of several years. The strategies or plans farmers pursue are clearly not isolated from the land they manage, and are indeed, intimately connected with it. It is a region's history (especially the forms of land use, ownership structure and management), and its political, economic and cultural context, that in turn shape the full picture of the situation in rural areas and in agriculture in each given region.

In the case of Latin America, from the mid-20th century onwards a major role in the processes supporting farm production was played by the policies adopted in given states (e.g., in some cases more or less successful attempts to introduce farm reforms), programmes for the agricultural colonisation of new lands like *SUDAM* (*Superintendência do Desenvolvimento da Amazônia*) and *Sudene* (*Superintendência do Desenvolvimento do Nordeste w Brazylii*) and ultimately programmes of technical assistance for farmers, especially involving the provision of higher-quality seed and plant protection agents (partially resulting from the "Green Revolution" programme introduced in Mexico, amongst other countries).

A given region's story as regards the ownership or stewardship of land, forms of land use and the relationship between the large and small producers of food all provided for the shaping of agricultural regions more or less well-integrated into global and domestic markets. The development of infrastructure (especially of roads) has also been very much linked up with historical processes and again has done much to encourage or hinder the contemporary development of production.

In line with views on the contemporary development of rural areas and agriculture in the countries of Latin America, economic situations need to be looked at from the point of view of one or other of three approaches:

- the theory of under-development, in which it is over-population, technology transfers and the diffusion of development that are emphasised;
- neo-Marxist theory regarding disparate development, which is focused on imperialism, dependency and global systems;
- an environmental concept as regards sustainable development (Garcés Jaramillo 2011: 42-43).

Irrespective of which approach is applied in analysing the situation, rural areas and agriculture in Latin America find themselves in, a view continuing to be adhered to holds that poverty reflects a lack of development and represents an element of under-development (Yapa 1993: 255).

Yapa disputes the traditional approaches referred to above, claiming that the main problem lies in the process of development as such, with "modern poverty" representing a form of shortage induced by development, given that the very bases underpinning subsistence or conditions for production are breaking down (Yapa 1993: 262). For example, where modern techniques or technologies are applied in agriculture, this leads to the destruction of existing resources, while generating a demand for new ones. Modernity also manifests itself in the introduction of new crop varieties more disease-resistant than their predecessors, and often modified genetically. However, locals are aware that they are losing the traditional kind of crop-growing they once engaged in, with no possibility of a return to it at some future date. What is more, in the Andes, the skills and knowledge as regards crops grown since the times of the Incas are being lost. Only some local producers seek to go back on to the market again.

The analysis of selected examples from the Peruvian Andes has confirmed the suitability of the concept of foci (elements representing intermediate stages) presented previously by Yapa, at least when it comes to the links in the chain ensuring the pursuit of a given kind of farm production in a given area, as opposed to any other kind. That author states in the context that farm production depends on the intensity and quality of a system of technical, social, environmental, cultural and academic relations distorted by ideas and visions owing to a reductionist approach (Yapa 1992: 256). The relations in question should not be understood as a separate, analytical category, but rather as dialectic, given the way that they are ongoing and constantly interacting, in this way maintaining the dynamic of the process (*ibid.*). Analysis does not allow for the presentation or even outlining of the networks that provide the linkages, hence the need to define the problem of poverty and to take a second look at the prevalent research approaches where that issue is concerned (Yapa 1992: 256).

The academic debate on the subject of poverty is indeed very wide-ranging, with economic, social, cultural, political and other threads all taken into consideration. Work done earlier in Peru (Czerny and Córdova 2014), which represents a point of entry into the research on sustainable development presented here, was focused less on poverty as such, and more on the safeguarding of existence through livelihoods. Study of the latter issue in turn links up with many other similar approaches, including, for example, to quality of life and matters of wellbeing.

Issues of quality of life became the subject of scientific (including geographical) analysis from the mid-1960s onwards. From the outset, controversy was aroused, given the lack of a unified definition of the phenomenon, or else clearly-defined research methodologies. The improvement in living conditions achieved post-War in Western countries was *inter alia* manifested in raised housing standards, better nutrition, universal healthcare and increased amounts of time off from work. However, in summing up these unquestionably reasonable parameters by which to confirm improved living standards, Kałamucka noted that: "Steadily rising indicators for wellbeing have nevertheless become counterbalanced by serious threats, and most especially by:

- the degradation of the environment and exhaustion of non-renewable natural resources,
- the biological and psychological threats associated with limits to human adaptation to more rapid civilisational change,
- the over-development of organisational structures, with concentration and centralisation, not only in the production sphere, but also in ways by which control is exerted over society,
- threats associated with a crisis of values raised in industrial society, in association with excessive consumption, the pursuit of economic growth and wellbeing, and the growing omnipotence of science,
- the glorification of mass culture, subordination of art and entertainment to the media and transformation of the latter into instruments of manipulation (through advertising and propaganda)." (Kałamucka 2001: chapter 2, p. 16).

The author in question engaged in analysis of quality of life as defined in a variety of different ways; the work making it clear how imprecise the term is, and how it cannot in fact be made uniform on the global scale. This reflects the basic fact that, for example, the inhabitants of Warsaw see quality of life differently from people living in the Peruvian village where the author of the present study conducted interviews. Among the definitions cited by Kałamucka, several at least would seem applicable to research in the Andean states (Kałamucka 2001: chapter 2, p. 23). Among others, Wilkening (after Kałamucką 2001: 23) notes that human quality of life is associated with the means of functioning in the natural and social environment (1973); while Hornback and Shaw (1973) claim that it is a function of objective conditions suiting a group of people, as well as of the subjective attitude maintained towards those conditions by the people in that given group (after Kałamucka 2001: 23). Finally, there is the definition from Wallis (1976), in line with which quality of life entails a defined structure of values, plus the situation an individual maintains in his/her daily life over longer periods of time (after Kałamucka 2001: 23). The latter author concludes that: "disparities as regards the interpretation of quality of life as an academic category do not relate solely to the way in which the definition of this is determined. For there are also considerable differences of standpoint as regards the wider theoretical justifications and the proposed research methodologies, or even the basic conceptual framework on which studies are based. The limited degree to which the issue has yet been recognised from the geographical point of view justifies the reviewing of research concepts and achievements as regards the academic fields in which the output in this area is already considerable." (Kałamucka 2001: chapter 2, p. 24). Hence also the motivation for more-frequent work on standard of living – this being easier to measure using statistical methods (concerning income and expenditure). However, later work has made it clear that standard of living is a component part of quality of life that expresses the degree to which elementary needs as regards nutrition, clothing and housing are met. However, quality of life also encompasses such categories as living conditions (the nature and shape of the human environment) and style or way of life (Kałamucka 2001: chapter 2, p. 29).

The category of quality of life – while not that precise – represents just one of the more important indicators allowing us to study development processes, and the scope to sustainable management in rural areas of countries in the global "South", including in the Andes. Quality of life offers a hard-to-measure and not-always-appreciated research approach, especially where a subjective feeling of success and satisfaction with an existing situation fails to match up with the vision of wellbeing and the satisfying of needs typical for developed countries of the world. For, in the regions under study here, a harmonious configuration of family life and a balance of social development not always capable of being described and explained in terms of goods possessed, individual preferences or style of life constitute very important components of individuals' or families' own overall self-assessments. Quality of life perceived in this way is obviously therefore a category very largely subjective in nature, being determined in line with human needs, convictions and values, with a person's own assessment of their quality of life being dependent on systems of reference, including local political relationships and relations with the local authority. "Quality of life as a research category is holistic in nature, given that it links together social and psychological aspects, while not losing sight of material and ecological dimensions to life, although the weighting of the factors mentioned may differ in relation to the research standpoint." (Kałamucka 2001: chapter 2, p. 49).

In contrast, Otto Neurath is regarded as the creator of the concept of "living conditions", having brought the term into the scientific literature, postulating in 1917 that emphasis needed to put on a holistic conceptualisation of the description of living conditions for individuals and families (Lessman 2006: 5). In turn, Gerhard Weisser claimed that each individual has his/her own list of values and things sought and strived for, ensuring that universal values were not to be believed in (Lessman 2006: 7). In contrast, Nussbaum et al. do have faith in the existence of universal values that denote the trajectory followed by each human being, and the progress made (*ibidem*).

Ortrud Lessman notes similarities between the approaches taken by Otto Neurath and Amartya Sen (Lessman 2006: 5). The two authors she cites have similar views regarding inter-personal comparisons based around wellbeing and goods continuing to be present in space (e.g., nutrition is a function of what goods are on offer, but also generates wellbeing, or at least satisfies fundamental needs). The "quality of life" concept (originally related to the inhabitants of Vienna) is owing to Otto Neurath, while Amartya Sen brought to the research on development her "capability approach". Otto Neurath researches relationships between measurability and comparability, as well as promoting a holistic-type approach to regional studies, when he notes that life in every community comes under the influence of different types of law.

"Certainly, different kinds of laws can be distinguished from each other: for example, chemical, biological or sociological laws; however, it *cannot be said of a prediction of a concrete individual process that it depend on one definite kind of law only*. For example, whether a forest will burn down at a certain location on Earth depends as much on the weather as on whether human intervention takes place or not. This intervention, however, can only be predicted if one knows the laws of human behaviour. *That is, under certain circumstances, it must be possible to connect all kinds of laws with each other*. Therefore all laws, whether chemical, climatological or sociological, must be conceived as parts of a system, namely of unified science. (Neurath 1931/1983, 59, original italics.) (after the Stanford Encyclopedia of Philosophy 2010).

It is an approach of this kind that corresponds with the concept of research on underdevelopment and poverty adopted in this publication. Interviews with inhabitants of the Andes resident at various different localities showed these people to be full of satisfaction at the (actually very modest) improvements in living conditions in mountainous areas they have apparently experienced, with this opinion reflecting complex historical, social and political conditioning in the region.

Neurath emphasised that all the different assessments of the state and conditions in which people live arise out of comparisons made with our standards, and that the need would be to change the prism through which we define (or recognise) conditions in which this one way or another useful conditioning is generated or arises (after Lessman 2006: 4). The author in question defines his concept of living conditions as "a central concept that explains all circumstances directly and comparably conditioning (creating conditions for) the way of behaving of a person, his/her fears and joys – to have somewhere to sleep, something to eat, and on the other hand even fears as regards given diseases posing a threat" (Neurath 1931:40 after Lessman 2006:4).

Neurath's input into the academic discussion on ways of living made possible the further debate on different paths taken in life – and different ways of life – present in different cultures, under different kinds of geographical conditioning. His approach to living conditions can be said to have beaten the path to A. Sen's concept regarding the role of choice of possibilities. Sen explains that his idea regarding the latter is nothing more of less than a combination of different ways out (alternatives), that an individual may exercise, implement or be, i.e., in some senses the different roles (ways of functioning) that he or she may assume or achieve (Sen 1993: 54). Possibilities are in turn a combination of alternatives with different actions that may be achieved and may constitute a certain kind of set to be selected – from among a great many others – by an individual or a group (Sen 1993: 55-56).

So can the experiences of industrialised countries prove attractive enough for developing countries when it comes to the implementation of sustainable development strategies? The environmental currents in the former are concerned with economising, with the conservation of nature and with self-limitation (mainly as regards consumption, in the form of a specific niche where the attitudes of consumers are concerned). The basis here is an understanding that the environment has been recognised as of value in its own right. In contrast, in developing countries there is a prevalent view that valuable features of this kind can be better and more fully appreciated in the future, with some first period instead seeing intervention in the environment with a view to the resources essential for economic development being exploited.

In the countries of the global "South" there is a genuine rise in the level of awareness of harm inflicted – especially by mining – that leads to the destruction of the environment in which people must live. Nevertheless, there is all the time a widespread opinion that environment policy – and the measures taken to implement it – put limitations on local-level economic activity, and in that way conspire to leave part of the local community in poverty. Such opinions have also been encountered in Peru, by the author of this publication. In the urban-industrial systems of the countries of the North, there is an opposite situation whereby the environment for life has been confined and people strive to protect what is left. Ultimately, this shows that the process of economic development is associated with a shift in the priorities and values speaking mainly for "development" if this is able to take place in a "sustainable" way.

Nevertheless, it continues to be the case that intellectual debate is centred around two main approaches to development and the use of available resources in its pursuit. One view has it that nature sets barriers for human beings that the latter should not cross, while the other attaches greatest importance to safeguarding wellbeing and socioeconomic development irrespective of environmental limitations. Of major significance from the point of view of the relationship between humans and the environment is the fact that an appropriate level of sustainable development may only be achieved where there is management and utilisation of renewable resources, i.e., those that can be regenerated and reused. Renewables are thus considered to be those resources that can be created anew within a single human generation. However, given the very long periods over which geological change takes place, the term renewable would assume a still-wider scope. As early as in 1994, Binswanger was seeking to emphasise that for centuries (including in pre-industrial times), the economy developed almost solely through the utilisation of renewable resources, albeit with the precise process whereby exhausted resources regenerate not having been researched more precisely or observed in more than a cursory fashion.

We can seek to argue further on the basis of Binswanger's theory, if we for example look at the slash-and-burn agriculture practised by forest tribes. The latter burn an area of forest, using the field cleared of trees in this way to grow the plants they regard as essential to life. After just a couple of years, the soil is so nutrient-depleted that the tribes choose to move on elsewhere with the whole process starting again. After a period of something like 10-20 years, people return to the plot that they once made use of, burning the secondary vegetation as they go, in order that some crops might be put in. Progress in the case of the original tribes present illustrates that they have a good knowledge of the cycling whereby soils are renewed, knowing thanks to their observations when it will be worthwhile or necessary to return to places in which crops had grown in the past.

However, Binswanger (1994) claims that – among city-dwellers – the use of reclaimed biological resources has not represented any ecological problem with wastes (at most only a question of hygiene), on account of the relative ease with which these can be broken down. There has also been no development of a more precise awareness regarding the cycling of matter. It is for this reason that these processes have not been inscribed within economic thinking, with "nature" only in theory being included within the productive functions, or else not being treated in this way at all.

However, ecologists (Haber 1995) have proved unable to bring on to the agenda two phenomena of key importance to human-environmental relations, i.e., the rapid increase in the human population on Earth and the constantly-changing model of consumption of the inhabitants of industrialised countries, which imposes yet-further pressure on resources, and on the natural environment in general. What is more, ecological phenomena and processes are overlain by political relations that influence the choice (or lack of choice) of ways in which problems are to be resolved. Such problems requiring global, or at least regional, solutions include those of climate change and air pollution which go beyond national borders and require bilateral or multilateral decisionmaking; as well as the management of forest (especially tropical forest); water management; soil erosion and ongoing desertification and biodiversity.

Kopfmüller (1995: 107) further states that an economic approach to sustainable development demands a rethink, with this entailing the abandonment of certain basic ideas taken from Neo-Classical economic theory. An exclusive focus on the market paradigm, on efficiency and growth, or on microeconomic profits, is not compatible with the means of management anticipated for the future. Questions of distribution at the national and international levels arise here, and first and foremost a question concerning optimal levels of production and consumption in a given society that needs to be put to world society in the future (specifically a question as to "how much is enough?" Or better still: "how much is too much?").

In contrast, in line with the ecological approach, the central point for further discussion is the concept of the level of "Constant Natural Capital" (Kopfmüller (1995: 107)). It is in line with this that we obtain the principle that renewable resources should only be used at the rate at which their regeneration can proceed, while non-renewable resources should only be used at all to the extent that renewables are unable to take their place (*ibidem.*). In turn, harm done to the environment cannot go beyond the capacity of the ecosystem to regenerate itself. This then gives rise to the idea of threshold maxima for the degree to which a given natural system may be exploited (i.e., "carrying capacity").

Also of key importance is for discussions and operational programmes relating to implementation of the sustainable development concept to be explained and accounted for, with the concept of "social wellbeing" being expanded in scope. Finally, as regards the socio-cultural and political aspects, it is essential that participatory elements be put in place and/or enhanced, through the determination of appropriated institutional frameworks for action, along with the protection of cultural diversity and identity.

Chapter 4

SUSTAINABLE DEVELOPMENT AND POLITICS

Issues of the implementation of a sustainable development strategy in a defined region are also associated with non-economic factors, with the relationship as regards effectiveness of action between the spheres of politics, culture and society being crucial, along with the models of development selected. Politics in general, the political system, the political complexion of the governing party and finally the wider legal and political environment in particular are what determine the introduction or non-introduction of sustainable-development strategies as instruments, as well as their effectiveness and scope.

In line with the Brundtland Report, there is taken to be a list of principles, or even rules, that need to be heeded in the process by which resources are made use of. It is my contention that compliance with these is conditioned by what is understood broadly as the political model in the given country. Notwithstanding the policy declarations made at international fora, a lack of political will and/or instrumentation within a given country will make non-implementation a near-certainty. Beyond that, the application of the above principles again has to do with the already-discussed and often-sensitive issues of national sovereignty and neo-colonial policy. The courses of political boundaries often in reality fail to correspond with the ranges or extents of natural phenomena, which often operate on a supranational scale. Examples might be the movement of masses of air (e.g., those associated with hurricanes or gale-force winds), river courses (it is frequent for a river to run through many countries, to the extent that flood protection becomes a supranational matter), the pollution of the seas and oceans, the process of desertification (as exemplified in the Sahel zone of northern Africa, in which efforts to arrest the phenomenon involve many states working together), the exploitation of tropical forests and global effects thereof (harm to the environment and local people's living conditions, migration, infrastructural development, etc.). It is in this way that the implementation of sustainable development strategies and programmes becomes political in nature.

Equally, neither the existence of political will for – or the practical implementation of – sustainable-development principles and environmental protection programmes in a given country will be anything more than illusory if not matched by similar efforts in other countries, and above all neighbouring countries. Outlays of any country on the objectives in question are rarely adequate, so international assistance may well be needed. There are environmental relations between countries and regions that are as strong as those in other spheres, including the economic /political, the financial/political, the techn(olog)ical/political, and so on. Presumably, environmental relations between countries are also dependent on internal conditioning, e.g., the principle of national sovereignty when it comes to decisionmaking in respect of one's own country; the principle of non-interference and non-intervention even in crisis situations without prior agreement and consent, etc.

According to Huber (1995: 35), the key principles whose pursuit will facilitate development in line with the assumptions set out in the Brundtland Report may be stated as follows:

1. The development of humankind must go hand in hand with consensus as regards the capacity [carrying capacity] and generative power of an ecosystem [in other words its productive capacity or strength].

2. The degree to which elements of the environment and living things are burdened by emissions should not exceed their sensitivity [vulnerability] and their capacity to regenerate.
3. The level of consumption of renewable raw materials and energy (e.g., water, biomass and to some extent also soils) should not exceed the given possibilities for the ecosystem to regenerate or reproduce itself.
4. There is a need to minimise the level of consumption of non-renewable [exhaustible] resources (and what are involved here are not banal deposits like sand or rock, but rather ecologically-sensitive raw materials such as flat sites, as well as deposits of crude oil, coal and gas) through:
   - the substitution of non-renewable resources by renewable ones;
   - raised efficiency of use of mineral deposits and energy;
   - recycling, insofar as this makes sense environmentally and is justified economically [capable of being continued with].
5. Development and implementation not burdensome for the environment and "clean" resources, technology and new products serve to strengthen the efforts to achieve sustainable development." (Huber 1995: 35).

One result of the Earth Summit was to generate a permanent rise in the temperature of a debate that had performed a gradual disappearing act in the period between 1970 and 1992, or at least gone through a major change in frame of reference. The discourse in question is that concerning neo-colonialism and developmental disparities between the rich countries and those that until just a short time before had been colonies experiencing exploitation, primarily at the hands of European states. The economic exploitation of the countries of the global "South" had most often entailed the extraction of raw materials – more often than not in a manner not far removed from "pillaging", with little or no thought given to the environmental or social consequences. The resources in question were mainly exported to the rich countries, with a view to their supporting their dynamic economic development, first and foremost via ongoing rises in industrial output. Post-Rio there was something of a reactivation of the view that poor countries remained in thrall to rich ones. At the same time, a trend towards the better protection of the natural environment worldwide was interpreted by developing countries as meaning that – contrary to their own economic interests – rich countries would now be imposing demanding standards in order that their own citizens could be protected from the consequences of global environmental disaster. The policy was also treated as a threat to the development of the poor "South". As it turned out, such fears were unfounded, in that global corporate interests exploiting mineral raw materials dominate the economies, not only of the poor countries, but also of the rich (whose supplies of necessary raw materials they ensure), in the process destroying the natural environments of the former countries in a manner little short of criminal. There are also fears of the so-called "eco-dumping", whereby countries close their own markets to cheaper products (mainly agricultural) from the global "South". This scenario has in fact played out in practice, at least to some extent.

As the countries of the world differ in their levels of economic development and hence also in their political positions, questions regarding a neo-colonial policy on the part of the West, and prevalent economic relations based around dependence, continue to reappear as the discourse on sustainable development continues.

The relations in question relate primarily to the imposition on poor countries of the conditions in line with which trade is to take place. As suppliers of agricultural produce to their markets, the poor countries in question are dependent to an ever-greater extent on capricious economic conditions and consumer preferences in the developed world. The preferences and fashions in question are in turn shaped to a great extent by the marketing activity engaged in by transnational corporations. To ensure supplies of given goods, transnationals buy land in the developing countries, in order that soybeans, cotton, peanuts, bananas, citrus fruits, cocoa, flowers, vegetables and so on can be grown. They also ensure the subordination to their interests of local producers, which prove to be extremely vulnerable to the vagaries of trading conditions and to fluctuations in demand on the global market. It is not only transnationals that harm the environment, as domestic firms whose main operational goal is profit do just the same, in this way participating in an activity often referred to as quasi-colonialism (Schmidt-Bleek 1994: 145). The costs of destroying the environment are always far higher than those associated with its protection.

The UN documentation on sustainable development contains a great many data on the means and instruments brought in with a view to the intended objectives being achieved. In general, and after Hubert (1995), these can be divided into:

1. the international conventions requiring different countries to implement national programmes – and put in place legal frameworks – to ensure protection of the natural environment;
2. domestic frameworks regarding environmental protection that are to seek to harmonise resource management with sustainable development goals, as well as adapting it to international agreements and law in the formal sense;
3. economic instruments of environmental policy, especially those concerning payment for energy and other charges for using resources; as well as the assigning and assuming of responsibility for harm done to the environment;
4. regulations concerning environmental analyses and studies, *inter alia* as regards the capacity of the natural environment and its durability in the face of harm or excessive exploitation inflicted upon it;
5. provisions as regards public participation in consultations over the use of resources.

It may in general be thought that the process implementing sustainable development principles should embrace strategies for the relationship between civil society and the market economy, with this working to counterbalance excessively bureaucratic and centralised procedures, including those applying to the investments made by the large transnational corporations.

Chapter 5

SUSTAINABLE DEVELOPMENT, CULTURAL EVOLUTION AND CULTURAL HERITAGE

Heritage – as an element of culture – needs to be looked at from the point of view of its economic potential and the possibility of local and regional economies being created. As Immanuel Wallerstein had it, the cultural aspect to globalisation is a derivative of economic progress, the perception of which is closely linked to the stage of development the world system finds itself in (Kumar and Welz 2001). Going further, the role of heritage in the process of contemporary economic development needs to be analysed as one of the manifestations of the production *v* consumption dichotomy.

Humankind has been changing the landscape and the natural environment for centuries, with a view to ensuring increased production meeting needs, because consumption has also been going up. Constant, uninterrupted destruction of natural ecosystems has therefore been taking place, with the natural landscape transforming into a cultural agricultural landscape of which elements can at times resembled the natural counterpart (particularly in the cases of meadows and orchards, for example). However, it must be questioned how far the landscape can be regarded as a balanced system (and the creation of new farm production systems as a steady or persistent process of interference in the environment), given the way in which the task assigned to agriculture is to ensure the regular (annual or even more frequent) production of food. In addition, the features of the environment in which the production of food takes place are more and more often improved artificially (e.g., through the remodelling of irrigation systems, drainage, terracing and so on). This is therefore activity that modifies nature very markedly away from the conditions originally present. Effective land use in agriculture, construction and other economic activity leads to an increase in density of population, and thus to further pressure being imposed on land that has already been modified markedly.

It was the production of food that stimulated the expansion of cities for decades. From the point of view of their ecological features, the city and the countryside represent extreme opposites as systems. Indeed, it may be claimed that a new type of settlement ecosystem came into being, in the totally modified natural environment that the urban ecosystem represents. Unlike the also-anthropogenic ecosystem that the agrarian system represents, the city is a more-closed environment for life extremely transformed by human agency, being densely built-up, busy and noisy. Such urban ecosystems are not self-sustaining from the point of view of their functioning and further existence, for they require a constant influx of external elements essential to life, i.e., food, raw materials and energy carriers – as well as the securing of a continuous outflow (transfer) or utilisation of wastes, wastewaters and other residues not made use of by city-dwellers. In the face of such dependence on external relationships, can the urban ecosystem be sustainable in the ecological sense, or can the objectives inherent to sustainable development be achieved in it?

The agricultural and urban revolutions, albeit with a certain delay, occurred relatively evenly around – or encompassed – almost all cultures on Earth. This does not of course ignore the fact that particular individual social groups, or tribes, remained at hunter-gatherer level. However, it was only in the one cultural circle that we today call "Western industrial culture" that there came a third stage or step in cultural evolution entailing technology and industry. This was liberated (brought into operation) more than anything else by a transition from the "renewable resources" (notably wood) that had been the main source of energy for years to fossil fuels, which became relatively easy to transport thanks to the almost-simultaneous, and obviously interconnected, invention of steam power. In just a few decades a technological and industrial society emerged (took shape) – creating cities at the expense of land and much of what was left of nature. The latter were linked up by roads whose construction took naturally-valuable land to develop the transport routes.

So it was that there emerged an urban-industrial society more and more inclined to seek out wellbeing and comfort. This has denoted a demand for ever-better and more diversified resources of various different kinds. At this point it needs to be recalled that most of the resources utilised by humankind (including notably the raw materials) are distributed in a very diffuse, but also uneven, manner across the lithosphere. In an urban society, the understanding of the need to seek out and utilise resources represents one of the main objectives of human activity. In the Modern Era, the colonial conquests pursued by Europeans gave them the mastery of land on which there were resources they could use to further their development. In order to exploit these resources and exert control over flows of them, the colonial powers also pursued a policy of support for settlement of the colonies from Europe. Established in this way was a single global system of economic linkages augmented by geopolitical relations that allowed for rather free exploitation of some lands in the service of others, with little care taken to monitor or show concern for the quality of the natural and social environment in the regions that served as suppliers of raw-material resources. It would not be far wrong to suggest that the system referred to has continued in operation through to the present day, with the fundamental principles on which trade and mutual relations are based not having improved greatly. This denotes an answer inevitably in the negative to any question posed regarding the sustainability of the technological and industrial system of Western industrialised countries (created via global expansion into new areas and the first utilisation of areas hitherto difficult or impossible to access), as well as its attendant urban culture. Nevertheless, efforts are being made (if so far mainly pointwise in selected areas) to implement sustainable development principles more consistently. This is in particular true of cities in Western Europe.

Generalising on the idea of people being guided by sustainable development principles in the socio-cultural sphere, it is possible to cite the example of cultural tourism as an element seeking development in tandem with the retention in given places and regions of the features that have characterised them up to now, at least in the recent past.

Within geography, research into heritage is first and foremost associated with the geography of tourism. There can be no doubt that ever-greater interest in the development of tourism globally (and the role this plays in restructuring the economies of localities and regions, and perhaps even whole countries) has been ensuring the increasing importance attached to relevant scientific studies, but also planning practice, with the role of the factor of culture in the (sustainable) development of given places being stressed.

In 1991, Britton analysed the development of tourism in the wider context of contemporary cultural and economic phenomena. He found that groups and individuals create their image (and also build their identities) by way of the construction of a model for consumption that takes account of local practices and habits or customs. In this context, consumption comes to be a denoter of allegiance to a given social group (Britton 1991). Fashions for the consumption of particular goods and services are to be noted, with some of these fashions for a given product fading over time, while others appear. This therefore denotes a process of evolution in the culture of consumption. If interest in heritage on the part of different groups in society is to be perceived in these terms, it follows that this is also shaped by fashions and current expectations that are more and more in line with sustainable development principles.

Alongside the places typically recreational in outlook (and hence mainly offering visitors ideal conditions for rest and recreation), the offers available to tourists also concern places whose features are different, given that they mainly draw people in with their culture and heritage. The development of such places is being studied in the context of geography, with a link being stressed between the phenomenon in question and the internal potential available – this being less about the actual heritage itself, and more about a shaped sense of ties with the given place and with local or regional identity. In turn, the existence of attributes like local identity, specific features of place or inhabitants, a feeling of belonging and of being part of a community (and so on) are treated as elements capable of raising the level of utility of a given locality. The latter feature proves to be of particular importance when it comes to the development of tourism, and there is indeed a general acceptance that the existence of heritage raises the level of utility of a given place (Ateljevic 2000).

The utility of heritage when it comes to the development of tourism is linked closely with matters of the market. Several decades ago now, Britton (1991: 462) drew attention to the exceptionally important role advertising and marketing had to play in shaping popular perceptions of a given place. It was his view that this was an element by which the development of a locality is invested in, or – to put it another way – "tourist space" is invested in. A place is thus promoted by putting emphasis on its uniqueness and specific features, on the way in which it is embedded in local tradition, as well as the natural landscape. It is in this way that the marketing of heritage in the name of tourism should stress how consumers (i.e., visitors) acquire (through their visits) a product that is one of a kind, unrepeatable, and restored or reconstructed as necessary especially for them (Ateljevic 2000). In line with the views widespread in the social sciences, the logic and rationality characterising the production of goods is here transferred to the "free-time sphere", with huge opportunities in fact created for controlled and also manipulated mass consumption, with attendant creation of places and assisted development only taking place where the marketing is effective (Ateljevic 2000). Representatives of the "Frankfurt School" have termed the phenomenon of the creation of an economic sector that focuses all of its activity on the management of people's free time "cultural industry" (*industria cultural*) (Ateljevic 2000). The normal laws of the market and requirement that capital be accumulated apply here, ensuring a steady stream of new ideas on how available resources can be used and managed to serve the purposes of tourism (and generally a proposal as to how free time may be spent). At the same time, demand is stimulated, with this not only extending but also diversifying the heritage offer available on the market.

In seeking out new products for the market, in line with a strategy, the *industria cultural* makes ever-fuller use of heritage already recognised and catalogued. Tourism thus treats heritage as a specific kind of production system into which there is continuous further "inclusion of little-known or unconventional culture, people, places, behaviours and sceneries" (Britten 1991: 454).

Thus if heritage is "częścią przeszłości, którą wybieramy w teraźniejszości dla osiągnięcia współcześnie określonego celu" (Graham, Ashworth and Tunbridge 2000), the process of constantly offering the people new items identified as heritage may continue for a long time yet. The search for relevant objects, and simultaneous excessive use of those already found and made accessible to tourism can only lead to the transformation of that heritage, in the worst cases even ensuring its devastation and destruction. Besides that, it is the consumer (on the one hand) and the entrepreneur (on the other) who use heritage in accordance with ideas, values and needs of a contemporary nature. A hundred years ago, when the level of development differed from today's, there was a different way of using heritage to achieve economic objectives (for example the fashion for antiquity and visiting Egypt assumed a form quite different from what is seen nowadays).

Heritage is thus appearing once again as a resource utilisable in the processes of local or regional development. According to Britton (1991: 454), this resource gains incorporation within the tourism production system. Here manipulation with a view to the greatest possible profits being gained results in the creation of a new place that is different from that which existed prior to inclusion as a tourist attraction, with justifiable sustainable development often being set to one side in the process. The phenomenon whereby new places are created – or else existing ones very much modified – as a result of intensive tourist exploitation is dubbed "selling places" in the literature (Ateljevic 2000), or else – in Harvey's words: "speculative construction of places" (Harvey 1989, 1993).

Jackson and Thrift (1995: 205) seek to stress that, when given goods (products) encroach beyond their normal function and assume cultural or symbolic significance – as opposed to merely economic significance – the relationship between production and consumption becomes blurred. It then becomes possible to speak of an endless, mutually-reinforcing cycling of the two elements, with heritage being used to generate profits. The phenomenon is particularly observable in the global "South", where there is a false overstating of the role tourism is playing in development, with simultaneous disappearance of genuine indigenous elements of culture.

The consumption of heritage may thus be treated (and understood) as a socially-constructed activity defined (and often modified) by the elite, and accepted, exploited and ultimately consumed by the masses. For the consumption of heritage to take place it is necessary for this process to be sanctioned by society/the public, which is to say that there should be legal regulations allowing for it, as well as institutions supervising and enforcing the process. Customs, habits and binding principles present in the given society must also allow for the act of consumption, and finally it may not conflict with the ideology (religion) holding sway, and the values adhered to (such that, for example, Mecca, as a holy city for Islam, is not accessible to non-believers), and so on.

Where references to heritage denote simultaneously references to a place, it needs to be recalled that the latter can be perceived as a real object, but also as a symbolic object. In both cases, different actors in the economy make use of it to create a defined image of the heritage in question. This is achieved via the media, tourist agencies, local authorities, the intellectual elites, etc. These all construct their own iconography of the landscape (or place) which the mass of consumers then come to regard as something obvious – even when only perceived by them (Ateljevic 2000).

Another thrust to research that can be regarded as a culture-related approach to sustainable development is concerned with the strategies that the societies in given countries should accept in order for an ideal model of development that is ecological, harmonious and just to be achieved. It is most frequent for the subject literature to follow Huber (1995: 39) in mentioning three bases – or to put it in other words three strategies – for action, entailing sufficiency (*Suffizienz*), efficiency (*Effizienz*) and consistency (*Konsistenz*).

Where sufficiency (or self-sufficiency) are concerned, the question coming back again and again is that first posed by Swedish researchers, who in the 1970s initiated a wide-ranging discussion on growth, and on the way out of poverty, in the case of underdeveloped countries. Specifically, the question asked was: "how much will be enough?" (after Huber 1995: 38). The question remains as apposite today as it was then, notwithstanding that now being some half a century ago. The question raises a fundamental cultural question as to who ought to limit their consumption and striving for ever higher living standards – the poor or the rich? Might the then-popular slogan *living poor with style* (*ibidem.*) serve as the economic coercion of some eco-dictatorship, or rather as an appeal to limit consumption? The latter is something impossible in the era of global economic linkage. The global utilitarianism offensive and endless seeking out of happiness understood as the reaching of a standard of living already achieved by the rich countries represent the main definer of development for most countries of the global "South". Those who would think seriously about the adoption of a strategy for some kind of self-sufficiency of future generations would need to arrest the growth of population to the point where the number of people on Earth would fall back to the level present before the Industrial Revolution (Huber 1995: 40).

From the beginning of the 21st century onwards, the dramatic acceleration of rates of GDP growth in many countries of the global "South" – as a reflection of political change and the elite's greater-than-ever focus on growth, effectively at any cost – has raised more and more controversy in relation to the idea to link the economy with ecology. This is a trend at odds with the one to be observed in many developed European countries, in which – for several years now – there has been a prevalence of government effort seeking to usher in more and more new environmental or ecological solutions, as well as those that would prevent the further destruction of the environment and reduction in the level of resources (Bluszcz, Inderberg and Zerka 2015). The authors referred to here present the interesting experiences of Finland and Germany, wherein there is effective implementation of programmes for comprehensive sustainable development, with the latter being among the main political objectives of current governments (*ibidem.*).

In these countries there is successful application of an efficiency strategy whose task is the consistent bringing into force of practical and feasible principles of management in the given ecosystem. The programmes of job creation (especially in industry) that are planned and supported by development policy ought to be implemented with the most limited possible use of energy and materials (resources). A steady growth in the efficiency achieved by labour and capital is here augmented by an efficiency of resource-use set at an appropriate level. The sense of increased effectiveness and especially efficiency in the context of sustainable development lies in the fact that this represents an effort to minimise the use/consumption of resources and the burdening of the environment, as perceived in both absolute and relative terms. The means and paths leading to this entail improvement of the techniques and technologies employed in daily activity (e.g., more efficient engines, the building of incinerators, and so on), as well as in recycling and the cascade generation of materials within the framework of closed cycling. In line with a strategy of this kind, materials should be used for so long as wastes return to the natural cycle.

Hence, advocates of the implementation of sustainable development principles promote the longevity concept where goods for consumption are concerned. However, today we know that this conflicts or is indeed incompatible with the strategies of the large corporations, who wish to increase the turnover of goods no matter what the cost (hence also the claims that goods are made to break down or go wrong or go out of date, in order to ensure that the customer will soon be buying again). The efficiency strategy is the one most often (effectively always) adapted in economic activity, but the question left open for the economy seen in this light is how that can be linked up with the principles of sustainable development. Conditions in the environment that vary greatly from one place to another ensure that the introduction of the principles in question encounters serious limitations. The more that renewables are used in a given place, the more it becomes possible that efficiency can become linked up with sustainability (with much emphasis put on the renewability of resources).

Finally, cohesion relates to the properties of the given resource. The cohesion strategy employed in sustainable development models denotes environment-friendly utilisation and management, and care for the due and unencumbered exchange of energy and matter in the given place (Huber 1995: 41). This is to say that the functioning of the environment and the functioning of human beings in the given environment do not come into conflict or obstruct each other. Indeed, quite the opposite happens as relationships between these components become strong and mutually reinforcing. A cohesion strategy is in line with the objectives and principles of integrated protection of the environment (as distinct from an "end-of-pipe" solution where the protection of the environment is concerned). The philosophy here entails integration of anthropogenic (production- and consumption-related) changes with the transformation ongoing in the natural environment (Ayers and Simonis 1994). This does not represent a solution in all circumstances, for adoption requires precise determination of objectives, as well as measures and indicators by which it becomes possible to measure the degree to which these said objectives are being attained. The limits to growth, self-sufficiency and economic and environmental cohesion will all represent key problems for development policy for as long as people live on and manage the Earth. It is also important that the limits should arise out of processes of innovation, technological development and the creativity of societies in particular countries. Otherwise, they may be exceeded, with the effect that conditions for life on the global scale are made worse.

Chapter 6

A Geographical Look at Resource-Management Policy

From the earliest times, economic activity denoted the exploitation of natural resources, be these in the nature of minerals, or other kinds of product nature had to offer. The interrelationship between the bounty of nature in any given area (and environment for human existence) and the appropriate and unique ways in which people in that particular place have made use of these resources is what ensures that considerations of environment policy – including policy on raw materials – need to be holistic (treating the system of relationships between the environment and human economic and management activity as a single, integral whole).

Where geographical analyses are concerned, the assumption arising out of the sustainable development concept (conceived as a striving to ensure that the needs and present future inhabitants of the Earth are both met) throws into sharpest relief those conditions of the natural environment as can serve as a basis for the life of a community, and for an economy, that is in essence independent of the era in which the people concerned happen to live. Only with this kind of approach is a steady process of development possible, with the functions the environment is to serve in that process being safeguarded for many years to come, and in essence for ever. Held and Geißler claim that a key role in human economic activity on planet

Earth, as well as in the maintaining of ecosystem structure, is played by the temporal and spatial dimension to the said activities (Held and Geißler 1993).

In the face of challenges and conceptual assumptions regarding sustainable development that are expressed in the above way, an objective of any policy for the rational use of resources, most especially raw materials, is the devising and introduction of pro-environmental forms of management, as well as changes in the structure of the "basket" of resources essential for development. The trend to development most anticipated by greens would entail a reduction of the pressure to exploit resources, with an attendant genuine reduction in demand for the said resources. While analyses of global economic trends do indeed suggest a change in resource-use structure in recent years (especially where mineral raw materials are concerned), it would be hard to speak of any real reduction in the amounts supplied (Lim and Spanger-Siegfried 2004). The change in resource-use structure is in fact seen to result primarily from technological progress worldwide. The search for new materials, products and technologies *inter alia* results in changes in economic processes, while also influencing social development. It is thus possible to speak of a social environmental impact manifesting itself in lifestyle changes, not least involving different products selected, greater environmental awareness, and so on. However, none of this changes the fact that new discoveries and products alike require new materials, with the search for and extraction of the latter leading – or capable of leading – to the further devastation of ecosystems, with overall harm done to the environment. This is all the case with, for example, the prospecting for and exploitation of rare-earth resources present in developing countries.

In each case, be it at national or regional level, the approach to resource management should be one in which an inter-sectoral conceptualisation prevails (as well as one that operates between ministries where the institutional circumstances are concerned). For an integrated policy and management process where resources are concerned allows, not only for more rational use, but also (and above all) for the introduction, utilisation and ultimate disposal of costly technologies.

In discussions concerning the proper management of resources and changes in the structure of the "basket" of resources that still remain at the disposal of humankind, the key player is the human being – with his or her consumption models, habits and dependence on products originating from the surrounding environment (be these food items, building materials, raw materials for making clothing and equipment for the household, etc.). It is assumed that the human being "settled" in a given environment makes conscious use of its resources, and treats them as a good essential to life. There is an associated process of assessment or evaluation of the resources existing in a person's surroundings (Czerny and Córdova 2014). In turn, the level of environmental stress arising out of the over-exploitation of local resources, and an awareness that the need to exploit may soon result in shortages or absences (not merely for future generations, but even for today's inhabitants, and one's own family) is linked with culture and tradition. The research carried out by the author in Latin America has many times made it clear that cultures (habits, traditions, adaptations) define the way in which needs as regards consumption, and the structure that consumption assumes, are shaped in the given place, and translate into ways of living, and ways of thinking about local resources. Schneidewind claims that, at the heart of the relationship between human beings and their natural environment (or more precisely the nature around them) there lie cultural processes and phenomena "inscribed" in the given locality, as well as processes of learning and the shaping of value models among inhabitants in which ecological aspects are only now making their way into people's consciousness (Schneidewind 1993 a). This view can be argued against, however, given that observations on how Andean inhabitants and those living along rivers in the Amazon Basin behave make it clear that environmental awareness has long been strong. Indeed the indigenous peoples have had this since the dawn of history, only making use of resources to a degree that allows the given environment to replenish them, in this way assuring communities of continuing access to what for them are the essentials of life.

State environmental policy should go far beyond the scope of the existing strategies and programmes that many governments (especially developing-country governments) have prepared and then introduced, though in fact leaving them in place on paper only, given the lack of either possibilities or political will for assumptions to actually be enforced among different actors in the economy. If the fundamental assumption here is the maintenance of the natural conditions that provide for human life and good health, while at the same time ensuring preservation of the bases for the economy and human existence, then the over-exploitative utilisation and "management" of natural resources taking place in many countries is certainly not operating in the interests of that assumption. The phenomena in question are widely observed in the countries of the global "South", notwithstanding the possession by many of these of a document dubbed a pro-environment and sustainable development policy. Examples from there show clearly how sustainable development may not be the sole preserve of state authorities. If remaining economic and social actors do not heed the relevant principles underpinning that development, or else downplay them (as for example happens with the exploitation of the mineral raw materials being prospected for around the world), then the true introduction of the key underpinning assumptions is a mere fiction. At the same time, there should be a major widening of competences to include the local authorities and bodies charged with real-life enforcement of the laws and legal provisions designed to usher in sustainable development in practice.

Chapter 7

Sustainable Development in Developing Countries

The processes of development in the global "South" have been subject to assessments and diagnoses different from those applied in the cases of countries that are developed economically. While it is true that the UN, World Bank and other international organisations employ universal measures (mainly GDP) in describing development, academics (and politicians) are well aware that these are not designed to offer a reliable picture of the situation in the highly-polarised, disparate and often (ethnically, socially and culturally) mosaic-like societies in each of the states of the global "South". Hence the multiplicity of (frequently non-cohesive) theoretical concepts making particular use of the theory of modernisation as applied to developing countries.

A separate question concerns the propagation and introduction in planning and social practice of sustainable development principles. In the 1970s, as countries in Western Europe rid themselves of the production sectors polluting the environment, they most often relocated them to what were then the countries of "the Third World". For many of the governments of the countries in question this was a manifestation of the modernisation process, given that there was a high pace of industrialisation in line with Western models. Only after at least two decades had passed did we begin to see the opinions of local researchers and thinkers who had in fact protested against the practices in question. Hence also the widespread opinion that the cause of disparate levels of development lies in the improper use of natural resources, while environmental problems arise out of poverty. The achievement of sustainability (or sustainable development) is seen as possible only where the economic growth that takes place is capable of overcoming poverty.

Opinions of this kind were still widespread in the 1970s, but today they attract fewer adherents, with the discussion on the sustainable development of peripheral regions have moved on to a focus on the possible reconciling of objectives set by environmentalists on the one hand, and the interests of local development on the other.

As different authors (including Dietz and van der Straaten 1992 and Renn 1994) have noted, sustainable development is a normative and normative/ethical principle proving even more attractive than that of social justice. It links together economic development with recognition of a certain ecological capacity, and it ushers in hope that conflicts between economic expansion and environmental limits can be quelled.

It is quite well-known that "sustainable development" is regarded as a concept by which to overcome or at least ease glaring economic and social disparities between the industrialised and the developing countries, and this not only today, but also in the future (on the basis of the so-called "intra-generational equity"). The assumption here holds that the poverty and constant shortages facing a majority within the society of each developing country favour an over-exploitative, chaotic and non-considered utilisation of resources. The over-exploitation showing little or no care for the natural environment is pursued by individuals (e.g., as trees are cut to serve as fuelwood), by small-scale entrepreneurs (e.g., as small, often family-run firms engage in the collection and sale of the eggs of turtles) and by large concerns that destroy the environment with impunity (e.g., by flooding the tropical forest with crude oil flowing from improperly-sealed wells).

These activities and a great many more are tolerated by those in authority for a variety of reasons – because auditing and monitoring measures are weak, on account of corruption, because people's basic needs are not even being met by the state institutions responsible, and so on. The

Brundtland Report also showed how the problem of the destruction and improper management and use of resources, as well as their immoderate exploitation, combine with a lack of programmes to protect nature effectively in constituting a constantly-present element of the North-South relationship (Haber 1995). For such reasons, the debate over sustainable development from the outset spilled out beyond the framework set by ecology, to take in a wide variety of issues on which relations between nature and the human beings utilising it are seen to depend. Thinking on sustainable development also took in new themes that were initially confined to land management (spatial planning and physical development), but then later (mainly) took in issues of cultural and social evolution, as well as the influence these exert in the way the environment is perceived, and the approaches adopted in its management.

As in the industrialised countries, so also in the developing ones, the bringing into effect of sustainable development assumptions has been linked first and foremost with an increase in the numbers and sizes of protective zones. Such activity goes hand in hand with the sustainable use of resources and the development of space in accordance with a strategy for sustainable development. However, effectiveness here is much greater in economically-developed countries, which invest large sums in "ecodevelopment", to put it in general terms. The process advances only much more slowly in the developing world, though the rate and scope of introduction of relevant principles in fact varies greatly from one country of the global "South" to another.

On the one hand, the structuring of the landscape and distribution of forms by which it is used require a clear formulation of objectives. And there can be no doubt that a key example of such a goal is to live in an undestroyed, uncontaminated environment that is also able to make a positive aesthetic impression. However, when it meets up with matters of the use of land and landscape, the achievement of this objective encounters major difficulties. Only very rarely is a landscape developed and built up in line with some plan agreed to earlier. Rather, landscape are the products arising out of the complex interrelationships and often-contradictory interests characterising the many different users of a given area of land.

Chapter 8

The Management of Resources: Resource Management

The concept of resource(s) management, often implemented as raw-materials management, has ceased to play the role of static instrument, having transformed into dynamic means of reacting and making decisions in association with the exploitation, trade in and processing of the said raw materials, as well as the management of the wastes generated. This kind of assumption underpinning the new approach to resource management assumes that the objectives of development will be taken account of at all levels and all stages of product development (Henseling 1995). Emphasis is put on management of the cycling or raw materials (or resources), with this denoting precise monitoring and directing of the route that a raw material takes from the moment it is first obtained, via different levels of processing through to a final stage, which is to say the obtainment of product from the given raw material and the means in which it is used, as well as possible processes whereby it can be recycled and reused as a secondary raw material, as well as disposed of (if not used again) (*ibidem.*). The concept of managing the life cycle through which raw materials pass puts emphasis on elements of modern environmental policy that are made possible by technological process, while at the time underlining the role in the process of the social actor (via public participation).

Management of the cycling of raw materials seeks the kind of shaping of the process by which the natural environment is exploited that will allow the factors of production that natural resources represent to serve humankind for as a long as possible. This must also be linked with the achievement of the economic, social and environmental goals that politicians and entrepreneurs have set. This in turn means that the process must bring in a wide variety of actors making direct or indirect use of environmental resources, which is to say entrepreneurs, consumers, the organisations of different types operating within civil society and ultimately the state itself, which in many cases owns at least some of the resources in question.

The main actors managing the cycling of resources are the state and entrepreneurs. The role of the enterprise is to give effect to principles regarding the rational and economical use of resources, while the task of the state is to ensure that all entities in the economy have appropriate frameworks for their activity, by way of formulation of the key objectives where the management of resources is concerned. In this respect, it is of particular importance that external influences and requirements be adjusted to conditions within the given state. State-level tasks also include the identification of the branches of the economy critical to the achievement of economic and political goals based around the utilisation of resources.

Alongside the two key actors managing the cycling of raw materials already referred to, i.e., the state and the economy (or more precisely the enterprises operating on a country's territory), there are many other actors exerting an influence on resource cycling. These include consumers, trade unions and environmental NGOs. Through the demand exerted for "environmentally friendly" products (e.g., those whose production does not necessitate exploitation of resources, and whose disposal does not increase amounts of waste going to landfill), consumers can reinforce economic decisions determining directions to further development. In turn, organisations can engage in educational activity that backs appropriate, pro-environmental behaviour in society. The roles of unions and employees' organisations may be similar.

The dominant role of the actors referred to above where resource cycling is concerned makes it particularly important that cooperation and exchanges of information between them be developed successfully. The need for the different entities in a given chain to interact can be achieved by way of joint analyses, trading, support for innovation, and so on.

Discussions on the wise or responsible use of resources and protection of the natural environment bring together the interests of ecologists or environmentalists, politicians and investors. Ecologists draw attention to the fact that the Earth is the environment for human life, with our quality of life capable of being influenced greatly by the way in which we elect to use resources. The significance of proper stewardship of the Earth and management of its resources in line with sustainable development models or strategies is often made light of, and underappreciated. On the Earth's surface, the changes ongoing are ones brought about by human activity, resulting in the generation of clearly definable new anthropogenic "landscapes" (Müller 1977, Haber 1995). The landscapes shaped by human activity are also durable, though they will not persist in the same form unless human interference is maintained all the time. Thus, for example, natural systems converted into agricultural ones (with fields, pastures, gardens and so on) continue to have much in common with what was present before they were established, but they will not prove possible to sustain if human activity is discontinued (*ibidem*). People have also proved capable of creating an entirely artificial ecosystem mainly comprising artefacts (built items of all types), with these being formed from areas of continuous settlement, or else the linear transport networks introduced into the structure of the landscape. The cultural landscape arises out of the natural landscape, and in given cases is itself natural (or better quasi-natural), consisting of agricultural plus forest ecosystems, as well as urban and industrial ones. The latter exist at the expense of the former, with the agricultural/forest ecosystems in turn present at the expense of natural ones, first and foremost of forest. As a result of all the activity seeking to reshape the natural landscape, some 40% of the land surface originally occupied by natural vegetation is now in use by human beings (Vitousek et al. 1986).

More than 100 years ago, the first attempts were made to manage (or regulate) the exploitation of resources, in particular through the protection of the environment from over-expansive use. The establishment of protected areas, i.e., National Parks, Nature Reserves and other areal forms of nature protection has had as its aim the prevention of disappearance where natural ecosystems are concerned (or at least a deceleration in the rate at which they were or are being lost or damaged). In the meantime, some attempts have also been made to reconstruct ecosystems destroyed at an earlier date, thanks to excessively intensive exploitation, manifested for example in the salinification of soils, the chemical contamination or ground or surface waters, and the pollution of the atmosphere.

However, it has not proved possible to hold back or contain the processes of urbanisation and urban sprawl, with the result that construction of the urban type – and the industrial, service-related and other functions associated with it – have continued to spread on to ever-greater areas of land. Equally, more recent years have seen efforts made to amend urban plans for physical development, which are now to reflect the new preferences of inhabitants of agglomerations and a model of consumption that entails a return to the compact city form with the arrangement of both horizontal and vertical urban green space.

The instruments in spatial planning, regional planning, urban planning, development strategies and so on in practically every country in the world have as their task the monitoring of the state from the point of view of forms of land use and care for the creation of harmonious spatial structures. There are many examples showing that spatial development plans are really brought into force, but do not take account of elements of environmental protection or the protection of the existing landscape. Illustrating the process rather well is a study of Tipiquaya – in the metropolitan Cochabamba region of Bolivia (Cesar – Tipiquaya 2015).

The growing threat to the environment and harm done to it in recent decades illustrate the way in which risks are ignored by local authorities, as they issue permits for new developments without paying much attention to matters environmental, in a procedure that is quite clearly capable of generating corruption. At the same time it is environmental protection legislation itself – as well as landscape-planning instruments formulated appropriately in accordance with it – that represent the right tools for introducing sustainable development principles at local and regional levels. However, the opposition of different local actors which should be complying with the regulations implementing those principles is so strong in many countries that provisions remain "on paper" only.

Disparities in approaches to matters of environmental protection and management also reflect the fact that ecology and landscape ecology are disciplines that developed rather in isolation. In any case, planning made rather sparing use of knowledge on the functioning of ecosystems on the one hand, and on the evolutionary development of cultural landscapes on the other. Today likewise, they are not the subject of any more lively discussion, as conditions underpinning exploitation – and potential harm to the environment – in a given place are analysed. The uneven distribution of resources and environmental costs of their exploitation in a given place ensure that ecologists are once again studying the degree of vulnerability and persistence of a place in the face of particular economic activities (Haber 1995).

In the earlier stages of the development of the cultural landscape in Central Europe, the suitability of a place from the point of its being used by human beings was of particular importance, and emphasised by the experience of the then inhabitants. As Haber (1995) shows, it was in just this way that agriculture developed on deep fertile soils that had developed from loess, or else from other kinds of sedimentary rock. Shallow, poor soils remained overgrown with forest or gave way to pastureland. In forests growing on soils not suitable for cultivation, the excessively intensive exploitation of timber (for fuel or construction) also led to rapid degradation. Villages and towns grew up in places where there was access to water and those fertile soils on which agriculture developed. The transport of all kinds of goods, including of food to towns, took place by water. Nature's non-uniform distribution of resources, as well as humankind's main use of renewable resources, favoured the emergence of a spatially-diversified, non-regular, mosaic-like use of land and corresponding cultural landscape.

With the advent of the industrial era, this kind of attachment to places as sources of resources upon which development was dependent – as well as the resultant regional autarchy – ceased to be a feature inherent to the choice of space for economic activity and the development of settlement. New spatial structures appeared, as well as new scales of economic activity, and these were seen to be (more) independent of the particular locality. The rationalisation of the use of resources by humankind owing to modern techniques and technologies of production, as well as in construction, transport and communications, led to new forms of spatial management, but also to a unification (i.e., increased homogeneity) in the case of spatial organisation, as well as to a further curtailment of the area still valuable from the natural point of view. It is not only industry, but also modern, large-scale agriculture targeted at export (e.g., of soybeans) that can lead to the disappearance of whole ecosystems, or at least to far-reaching changes in them. An example might be the disappearance of natural (mainly forest) environments in SE Bolivia, with their place being taken by fields of soybeans, cotton and other crops valuable from the commercial point of view. Comparison of land-cover maps between now and 30 years ago makes the seriousness of the problem, and the real-life consequences of globalisation processes, abundantly clear.

Thus, on the one hand, knowledge on the level of destruction – and need to protect – the natural environment has been growing, with the effect – in the EU for example – that ever more initiatives designed to protect ever larger areas have made their appearance (not least *Natura 2000*).

On the other hand, in the case of the developing countries, the knowledge in question is also broad and widespread, but economic interests (and particularly those of the ruling elites) ensure that, by making decisions that assign concessions to multinational concerns, they are permitting outsiders to make huge further changes in the environment domestically. It is not merely that the process by which resources are exploited can bring about huge damage (as in Peru, where the desire to mine gold from the earth ensured the pollution of many rivers by mercury;

and in Ecuador, where part of the Amazon Rainforest was flooded by crude oil from the Sanshi area, while the River Montego became polluted with heavy metals, having been the main source of water in a fertile valley since the times of the Incas, etc.). For there is also the fact that the issuing of concessions to exploit metal ores – for example – implies conscious acceptance by the authorities of serious – most often irreversible – change in the natural environment (as when a mining investment by a Chinese firm in Peru led to the total disappearance of certain features of the local relief). Once made in the environment – as when the summits of mountains are levelled, tropical forest is felled, soils are polluted with heavy metals, and so on – changes can be either irreversible or at least requiring a very long period to elapse before some kind of reappearance in a secondary form can take place. The presence of a global market and of competition for resources leads to a curtailment of the natural landscape and ecosystem on Earth, with effects beyond the widely-cited global climatic warning that are simply unknown at this stage. The economic growth in the countries of the global "South" together with increased incomes and an improvement in living conditions for local inhabitants helps to create new jobs, and reduce disproportionalities and disparities in development opportunities between regions, etc. These kinds of achievements mean a great deal to politicians, with the inevitable result that environmental protection will always take second place. Thus elements of sustainable development are only introduced into various kinds of activity at local or regional level to the extent that they do not stand in the way of growth processes measured by reference to hard economic indicators.

Chapter 9

PERIPHERAL REGIONS

The term "periphery" carries various connotations with it and can be understood in different ways depending on the function ascribed to thinking on it, and the way in which it is perceived or conceptualised. It is most frequent for geography to refer to the location of a given region in respect of the centre of a country, be that in the geometric, economic or political understanding of the term. "To be peripheral" thus relates to both distance and status. In each case, the terms periphery and peripheral entail a comparative element. By definition a periphery can only exist if there is also a centre somewhere else. The geographical connotations often also link up with matters psychological, in that "a region lying at the periphery of a country is poorly-developed" (Czerny, Czerny 2002) (Janicki .... Łopuszna) – a contention that need not be true, even if it is mostly imagined to be. It is thus clear that relationships between the centre and remaining (more or less peripheral) regions, or else between a more literally (geographically) defined centre and a periphery indicate unequivocally the situation a given area finds itself in. The concept of the centre and the relationships that make possible its definition have been written on at length by K. Handke (1993). According to that author: "the centre fulfils functions that we conceive of as central, while the region is situated somewhere beyond the relationships that are thought of in this case. Historically, what was met with more often was the contradistinction between the capital and the provinces or the capital and the periphery." (Handke 1993: 105).

Yi-Fu-Tuan in turn puts emphasis on the importance of words like "close" and "distant", which attest to relationships between people that also extend in the directions of friendship or hostility, as well as closeness in the geographical sense of familiarity with a given area (Yi-Fu-Tuan 1987).

In the view of Handke (1993), any schematic depiction of what is not the centre, and is thus the periphery, has as its components:

(a) horizontal spatial elements, i.e., a subordinate place in the system (at some distance from the centre or the zero point);
(b) vertical spatial elements, i.e., some position in a hierarchical system, always lower than the most elevated (below the top on the axis);
(c) evaluating elements, i.e., subordination to the centre, often extensive in spatial terms, but of lesser independence or entirely deprived of independence, and also with lesser authority and prestige;
(d) such linguistic elements as emphasise subordination (Handke 1993: 117).

"The very relationship of centre *versus* non-centre is universal in nature, because such a configuration is generated in every social space, if with the process of delimitation involving the content and nature of the main component parts [...]. In societal practice and looked at from the perspective of history, the system undergoes many and varied modifications, since the essence of the centre is based first and foremost on authority, strength, prestige and money…" (Handke 1993: 117).

Alongside markedly geographically relationships → being at the periphery and being peripheral, there are also cultural and psychological connotations influencing the way a given region or part of given territory is perceived. The term is associated with a presence on the margins of the main currents to economic political and social life, and hence with features of being weaker, not taken seriously or underrated.

Analyses carried out in relation to developing countries do not offer an unequivocal definition of what the periphery might be. In line with the original concept of the centre versus the periphery arising from the discussion on dependent development taking place among Latin American intellectuals and economists (with Raul Prebisch to the fore), the world has regions that are highly-developed economically (the centre) as well as regions that are only poorly-developed (and hence peripheral), but which do supply the former areas with their main raw materials (Rościszewski 1974). Analysing the situation Latin American countries find themselves in, Prebisch states that the underdevelopment of the region is structural in nature, reflecting circumstances first put in place in the colonial era, and entrenched from the 19th century onwards – i.e., from the time countries in this part of the world gained their independence and headed off along the path towards the diversification of international economic relations.

Among the countries whose positions as regards commerce with and investment in Latin America were highest were: the USA, the UK, France and (from the end of the 19th century) Italy and Germany. From that day to this, many countries of South America like Argentina, Brazil, Uruguay and Chile have continued to assign a key role to the export sector, whose structure continues to be dominated by raw materials and primary products, be these mineral, natural or agricultural. At most these are augmented with articles that have been processed to a limited degree (Prebisch 1959). The directions the economic development of peripheral countries has followed are thus ones subordinated to the demand exerted – and the strategies for development pursued – by countries of the "centre". This means that the centre-periphery relationship has been unfavourable for the countries forming the latter, from the point of view of the development and diversification of their economies.

However, from the point of view of further research into underdevelopment, the contribution made by geographers proved to be important, since these workers stressed that the centre-periphery scheme repeated itself and had become entrenched in the spatio-economic and social structure of South American countries (Czerny 1980). Of pioneering significance in this area was the work of Milton Santos, which inspired a discussion on internal disparities to levels of development (Santos 1971). As early as in 1974, M. Rościszewski wrote that "In the countries of the Third World, the greater part of the territory, inhabited by a majority of the population, would need to be assigned to peripheral space. Matters of the development of the countries under discussion here also mostly extend to a remodelling of socio-economic relations of just this peripheral space. From these points of view, research into the nature, functions and internal differentiation of peripheral space, and [...] the latter's linkages with central space is also of great importance [...]" (Rościszewski 1974: 13). In the Polish geographical literature of the 1980s, this train of thought was *inter alia* developed by M. Czerny, as she wrote many times of the spatial disparities that characterise development within given national territories (Czerny, 1980, 1985, 1986).

Since the 1980s, disparities in levels of development within given Latin American states have started to widen. Beginning to appear alongside what are unambiguously "central" regions – most often countries' capital cities – there are regions that have come within the orbit of world trade thanks to globalisation, their relationships with the external market in turn becoming stronger than those binding them within the country. Urban centres have also been involved, whether these be industrial or service-related, or offering their products on the global market and modernising rapidly the production process on farms almost entirely geared to the global market (for example through super-automation of the production of wine, meat and cheeses, fruit and vegetables designated either directly for the world market (in the case of avocadoes, apples, melons, pears, American blueberries, artichokes and so on), or else for processing in the factories of the large multinationals like Dole, Heinz, Del Monte and others). The appearance of such enclaves of modern agriculture differing from the large *latifundia* from earlier times that were more engaged in extensive than intensive agricultural production has only served to widen the gap between regions in which the farming is relatively up-to-date and those in which there is a continued prevalence of subsistence agriculture to meet farmers' own needs, with only limited use made of modern techniques and technologies, to the extent that the main thrusts to development have somehow passed them by altogether. Regions in which this kind of farm production holds sway may obviously be regarded as peripheral (Czerny and Córdova 2014).

Analysis of the socio-economic situations rural regions find themselves in ought to offer an answer to a question as to why they have remained peripheral. Of course, peripherality may also be ascribed to regions where mining is carried on, but perhaps to a declining extent; or where other kinds of industry from the Ford era were in place but have since collapsed.

However, it is the purpose of this article to focus on rural areas, and there the focus will remain. In fact, studies and opinions on the causes of underdevelopment in this sphere are seen to be exceptionally wide-ranging and multi-faceted (……). But here the author seeks to engage in the more systematic organisation of causes which have in her view helped determine the peripherality of rural areas of the Andes, most especially in Peru.

Chapter 10

The Peripherality of Rural Areas in the Andes

The development of rural areas in the Peruvian Andes is mainly linked with agricultural activity. Since pre-Columbian times, the management of land for farming has represented the greatest challenge for those inhabiting the region in question. The scarce resource that arable land inevitably represented here was subject to special protection. For example, settlements of indigenous people were located on the slopes, so as not to take up highly-cultivable flat land along valley bottoms. Up to the present time, it remains possible to discern traces of this very way of proceeding along the Urubamba Valley in Peru. The traditional agriculture we are dealing with over large parts of the Andes has developed in the region for centuries, with the existing natural environment of the mountains being made use of, such that local inhabitants are indeed able to adapt to conditions present, however difficult they may seem at times. With the appearance of the Spaniards in the region, transformations in agriculture were above all encouraged by changes in means of land ownership or tenure applied, also with the growing of many crops not known before in this region. The meeting of two cultures also gave rise to major transformations of the natural environment, with the Spaniards taking on land not cultivated earlier by the indigenous peoples.

Several decades of research and analyses devoted to Latin America allow the author to arrive at certain conclusions regarding underdevelopment, and especially the marginalisation and peripherality of rural areas. While the list is not exhaustive, and is subjective, it is generated on the basis of what has been read, as well as interviews carried out.

The causes of uneven development need to be looked for among factors of an environmental, political, cultural, social and finally also economic (or more correctly a technical) nature. Beyond that it is clear that it would be difficult to ascribe disparities to the impact of any one of these factors. There are a host of possible and actual interactions and feedbacks between the factors mentioned that ensure each region a unique image in general terms and as regards spatial management.

The natural environment offers opportunities for development, with conditions different at each location ensuring a diversity of forms of agricultural use around the world. The inhabitants of rural regions tend to make optimal use of the environmental conditions they are exposed to as they seek to meet their needs. From this point of view, the environment may not be regarded as a factor hindering development – for in those places where human beings make use of their environment there are by definition conditions in which food may be produced. Those rather few places that do not allow food to be obtained are uninhabited. In these circumstances it becomes paradoxical that the natural environment is the factor most often invoked as a primary obstacle to development, though there are of course many recognised reasons why this should be so. One of these concerns the fact that environmental conditions are subject to variability that reduces output and sometimes leads to outright hunger. The climate may vary, but so also may soil conditions, conditions as regards water resources and so on, these all being elements that can determine amounts of food produced in a very marked way. If issues arise in this sphere, the talk then is of conditions unfavourable for agriculture. Abrupt, unexpected and in fact unpredictable natural events ensure that conditions for human management can become altered even in an environment that is and has been occupied. Historical descriptions of situations of this kind characterising South America are in fact many in number, and relate to the times of the pre-Columbian civilisation (most notably the well-known fall of the Maya due to far-reaching change in an environment that had been managed rather intensively), as well as to modern-day climate change genuinely giving rise to the destruction of crops and places of habitation alike. In such ways does an apparently well-rooted socio-economic structure in a given area need to alter as a result of environmental change. However, the assigning of value to these changes is not something to be attempted here.

Nevertheless, there can be no doubt that the main causes of developmental disparities that characterise and have continued to maintain a mosaic of better- and worse-developed regions are of a political and cultural nature. Complex ethnic relations between indigenous peoples and incomers from Europe have since colonial times been giving rise to marginalisation, first of the said native peoples, and later of slaves from Africa and their offspring. Feudal-type relations were maintained into colonial times, and their non-reform in the newly-independent states arising in the 19th century has continued to ensure a place on the margins of socioeconomic life for the greater part of the population in what was previously Spanish America. Discrimination, usury and disdain for everything of indigenous origin were for decades markers of the relationships between different social groups in this part of the world. And it is a sad fact that these problems have come through to the present day in some areas. Regions inhabited by indigenous peoples thus remain poorly-developed, to the extent that (without state assistance, at least) they have no chance of joining the global market. Things might of course have been (or still be) different were the state to have been (or now be) interested in the sale of mineral resource concessions to international concerns (as in the case of the Conga Mine in Peru). However, the effect here is merely to worsen an already tragic situation faced by local people, with a further fall in agricultural output in areas encompassed by mining activity, as well as emigration to the cities of the rural population (this most often ending in the depopulation of regions in poverty by farmers deprived of the means they need to live, expelled from their land, or left with no alternative but to sell it).

Further causes of the maintenance of peripheral rural regions are a lack of structural change in agriculture – not only a lack of effective farm reforms (for which most countries were not prepared politically when some decided to embark upon them in the 1960s), but also a failure to create effective legal and financial mechanisms that would allow small farmers to increase farm sizes, join in with the commercial production of food, and modernise production. These matters arose repeatedly as field work was being carried out in the Andean states. Political clientelism, strong solidarity within groups in society (an origin in the same place, proximity of residence, affiliation with the same ethnic group, etc.), frequently occurring corruption and ongoing illiteracy (especially among women) all combine to ensure a lack of development impulses among the region's entire population. Some are even deliberately pushed to the margins of political and social life, in order that they might be left in no position to benefit from any possible economic attainments, should these eventually arise in the region.

Ultimately, there is nothing more than a small group of inhabitants of each region that are able to hold sway over each region from the economic point of view. In Latin America these are the descendants of creole families whose wealth arose from the mere fact of the ownership of huge areas of land, as well as the possession of large labour forces and access to raw materials and resources. When Latin America began to experience a modernisation associated with the development of capitalism, part of the old colonial elite was in a position to invest in the developing industry. At the same time, new investors from other European countries (beyond the old colonial metropolises) were able to locate their capital in the ports or capital cities of the Latin American states. Location policy was then a reflection of the link between the supplier and the external market. Whole areas of the different countries began to represent some kind of raw-materials-related, agricultural and livestock-rearing hinterland for the trade with Europe and North America. The capital cities and ports were in turn bridgeheads for foreign investors, who bought here the raw materials turned into manufactured goods in the factories of the North. Relationships between centres and peripheries thus took on the form of internal dependences and economic links between the main economic centre of the given country and its remaining parts. Thus, for example, 1970s, Mexico City concentrated more than 54% of Mexico's industrial output, and almost the same proportion of its entire industrial workforce (Czerny 1985).

A serious obstacle to the reduction of disparities in the level of economic development between regions was constituted by poorly-developed infrastructure, most especially roads. For example, in the mid 1970s, Bolivia had just a single hard-surfaced road some 300 km long, while in Colombia as recently as in the late 1970s a car could only be taken along two asphalt-covered roads running north-south (i.e., one that led south from Bogota as well as a little to the north, and a second that followed the Cauca Valley from Medellin in the north to the border with Ecuador), as well as one running east-west which linked Bogota with Cali. To this day there is no section of the Pan-American Highway that would link Panama with Colombia, and at the same time facilitate the flow of information, not so much about the resources and opportunities this region has to offer, as about the violence and unjust treatment meted out to indigenous peoples by the old and new political/military/economic elites that control the land and the access to all resources.

Chapter 11

THE PERIPHERALITY OF THE SIERRA DE PIURA FROM THE POINT OF VIEW OF ENVIRONMENTAL, POLITICAL, CULTURAL, SOCIAL AND ECONOMIC-INFRASTRUCTURAL FACTORS

The peripheral nature of a given study region manifests itself in terms of various features. In the case of the region under study here these are both physical features – a location on the margins of the national territory in northern Peru, and others of a social or economic nature.

In the case of indicators of social and economic development it is only possible to compare the existing situation in the area under study with values reported for the country as a whole. However, it needs to be recalled that many indicators at national level do not meet criteria for a high level of development and so might also be considered to represent features otherwise typical for peripheral areas.

One of the key indices where peripherality and marginalisation are concerned is the one relating to level of education attained. Given below are data obtained in the course of fieldwork as regards the level of illiteracy in the communities studies, as well as the proportions that have completed either primary or secondary education.

Table 1. Level of education in three regions of Frias District, 2012

Level of education achieved (%)			
Level of education	***SECTOR BAJO***	***SECTOR MEDIO***	***SECTOR ALTO***
Illiterate	3.17	9.21	6.2
Non-completed primary	30.2	32.26	22.6
Completed primary	22.22	20.01	17.4
Non-completed secondary	12.82	8.41	4.5
Completed secondary	8.05	8.91	9.23
Higher	4.05	8.1	8.46
No response	19.45	13.05	31.5

Source: questionnaire-based research, December 2012 in: Córdova-Aguilar, 2013:42

It results from this that more than half of the population in the area studied has education to primary level at best. Furthermore, the interviews carried out make it clear that those claiming to have non-completed primary education are often unable to read and write.

In the course of the interviews, local people complained about the quality of teaching – the lack of assistance, difficult conditions present in schools, and poor teachers with inadequate preparation to do their job. The three main postulates regarding an improvement in quality of life mentioned by respondents include a raising of the level of education.

According to data from Peru's statistical office (*Compendio Estadístico del Perú* 2011, vol. 1: 134), as of 2010 Frias District had one doctor's surgery and seven medical rooms. This number is insufficient to serve all the residents of the District's villages, all the more so when reaching one of the medical rooms may be a process taking several hours to achieve. The crisis of medical care is further deepened by the fact that these potential treatment points are lacking in both equipment and medicines.

In the circumstances of what is effectively a lack of medical care, diseases (including those of parasitic origin and with insect vectors) can spread effectively. There is a high mortality rate among infants and young children, as well as an under-nourishment problem among infants that is first and foremost the result of an inappropriate diet lacking in protein and vitamins. Average life expectancy in Frias District is 67, as compared with 69.4 in Piura region and 72 in Peru as a whole (*ibid.*).

From among the economic factors what above all attests to peripherality is a prevalance of subsistence agriculture over other forms of economic use. In these circumstances, the crop structure is dominated by food plants grown to meet families' own needs. A poor state of roads makes any commercialisation of agriculture difficult, such that the few attempts made to achieve this are the exception rather than the rule, with farmers needing also to overcome many challenges and obstacles of an institutional or bureaucratic nature if they wish to stay in business. In the course of discussions run with them, farmers complain about the role of "middle-men", who buy at times of a glut in production and often therefore offer prices that fail to cover production costs incurred.

Chapter 12

VIQUES DISTRICT

Viques District is in the *Altiplano*, in the southern part of Peru's Mantaro Valley, at altitudes between 3160 and 3800 m a.s.l. It covers some 3.7 km^2. Flat land is dominant here, with the highest altitudes reached in the southern part. The locality serving as the seat of the District authorities – which goes by the same name – is situated on alluvial terraces in the Valley and is just 14 km from the provincial capital of Huancayo. The District has a population of 2100, with people who are occupationally active accounting for 57% of this total (INEI 2010). The population pyramid depicts a prevalence of women. Peruvian statistics include a category "unremunerated work in the household", and in the case of Viques this category applies to 20% of the occupationally active population.

The river valley has fertile soils, so agriculture has been developed here for hundreds of years now. Even in the days of the Incas, the Mantaro Valley was among those areas serving as "granaries of the empire". Farmers in the Valley itself need not complain about any lack of water, as the lower terraces have many small bodies of water and wetlands, in which water is retained throughout the year. The situation is worse in the mountains, where periodic shortages of water do limit farm output.

Viques itself is reached by a canal bringing in water for agricultural purposes and representing an element of the valley irrigation system known as the *Irrigación de la Margen Izquierda del Río Mantaro* (CIMIRM). The system by which fields are irrigated was installed as early as in the times of the Incas, and parts of that have survived through to the present day. The mean annual temperature here is of 12.9° C, while the minimum is of 5° C. However, these are mountainous areas, so annual and even daily amplitudes of temperature can be very large. On average, some 730 mm of precipitation fall each year, though this reaches 1060 mm further up, at altitudes above 3500 m a.s.l. However, evapotranspiration is also at a high level, ensuring that farmers do indeed complain of water shortages as crops are growing. The rainy season here is in summer, i.e., the September-March period, and it is then that around 80% of the annual precipitation is recorded.

The traditional crop plants here include maize, potatoes, barley, wheat, flax, amaranth and beans. The list makes it clear that the diet of local people includes both products of South American origin and those brought in later by European colonists. The rather thick layer of soil present tends to remain moist, thus favouring crop-growing even at altitudes above 3000 m a.s.l. Particularly good yields are achieved from the growing of potatoes here, though some areas of soil are (now) poor in organic matter. (ZDJĘCIE UPRAWY ZIEMNIAKÓW)

However, the key factor conditioning the development of agriculture here relates to the weather conditions beyond 3000 m a. s. l., where potatoes are for example cultivated. Needless to say, yields here can be affected by the more frequent occurrence of ground frosts, and it is crops grown within the space of a year, with both irrigation water and fertiliser supplied, that do best in such conditions. Soils on the slopes are stony, supporting only a rather sparse plant cover (of grass, herbs and spiny shrubs). These areas are therefore devoted to sheep-grazing. As has been noted, the river valley also has wetland areas and swamps with pools, both of which support aquatic vegetation of importance to populations of birds. It is these areas that may now support a woodland of eucalyptus trees, and the latter are cut to supply wood for building and fuel to the local people (other parts of the District are lacking in woodland, which occurs to only a very limited extent anywhere in the Mantaro Valley). The areas with water are in part given over to a project designed to attract tourists that is described more fully in a further part of this chapter.

The environment in which local people live is diverse. The Mantaro Valley flood terraces have wetlands and even open water, in an area only utilised to a limited extent, but severely littered. Otherwise, it might be put to use in tourism. Not far from there is the area most attractive in terms of both utilisation and the actual incomes people manage to obtain, i.e., the rest-and-recreation-serving Mayupampa area, with its series of pools and other installations playing host to people coming out into the countryside in large numbers from Huancayo. Here at least both waste management and the protection of eucalyptus woodland (alien to the environment as it may be) can be considered appropriate. Finally, the higher land surrounding the town is used mainly in crop-growing, or else as pasture for those involved in raising sheep.

Chapter 13

THE ECONOMY

The Central Andes in Junín Province are rich in mineral resources (both metal and non-metal). Soils along the southern margin of the Mantaro Valley are very fertile and have been farmed intensively for a very long time now. For this reason, this was a densely-populated region even in the pre-Columbian era. Viques lies at the southern end of the Valley, and the whole District covers just some 457 ha. The fact that the entire Mantaro Valley continues to play host to a large concentration of indigenous peoples favours the retention of culture and traditions dating back to both the Spanish era and the pre-Columbian period. The folk culture and popular culture here is very rich indeed, while a diversity of forms ensures interest in – and the promotion throughout Peru of – this region's art. In the same way, it encourages the development of ecotourism and ethnic tourism alike.

It is possible to distinguish between four main categories of land use in Viques District as such: in the river valley, the so-called Mayupampa area serves recreational functions and has further potential to develop; above – on the second terrace – is where the more extensive built-up area of the small town is located, along with a railway line that is these days used only a couple of times a week; higher still, the slopes have fertile cultivated areas; while the summital parts (including the so-called *Vista Alegre*) are used in the grazing of livestock.

(TU ZDJĘCIA Z TYCH CZTERECH KRAJOBRAZÓW)

Within the aforementioned 457 ha area of the District as a whole, some 30.6% consists of farmland, though within this only 4.9% is irrigated (*Censo Nacional Agropecuario* 2012). Given the intensive use in agriculture, this area is almost entirely lacking in woodland (indeed the original forest here was felled in pre-Columbian times). There is a certain amount of artificial replanting with different species of eucalypt – obviously alien to this environment, though adapting to local conditions rapidly. Such trees are first and foremost planted in wetland areas of the lowest river terrace.

At altitudes this far above sea level, the traditional crop species was the potato, along with root crops (notably oka). These plants continue to be the dietary staples for inhabitants in the higher parts of the Andes. However, programmes supported by the local authorities and seeking to achieve a modernisation of agriculture have ensured an increase in the area under more permanent forms of cultivation, primarily with apple trees.

Local development is based around community organisations that seek to encourage the generation of surplus output capable of being sold on the regional market. However, a quite different kind of communal undertaking is the *Aquapark* project centred around lagoons in the Mantaro Valley. Work started on this at the end of the first decade of the new millennium, with the assumption being that at least two bathing pools would be put in place, along with a slide and other water installations – as well as the service facilities to go with them (changing rooms and showers, but also a cafeteria). In a true sense, the entire local community did become involved in project implementation, and the gastronomy is in the hands of the local women. Furthermore, the District authorities ensured it was in particular single women or those otherwise heading their families that were offered positions preparing and selling meals. At the present time, more than 70 women are holding down employment of this kind, ensuring the status of the park as a major source of income, given the small size of the population overall. Further beneficiaries of the development include local providers of transport in the form of both taxi drivers and rickshaw owners.

Farmers can count on obtaining better-quality maize seed for sowing. Programmes applied by NGOs also support the growing of the so-called “by the house” food crops (*huertas familiares*). They therefore supply young vegetable plants or certain sapling fruit trees (including berries). They then encourage farmers to use any surplus fruit or vegetables they may obtain in this way to sell at the town market in Huancayo. However, such extremely fragmented farm production does not favour development of a food industry, even if intensified production ensures that a few farmers are now able to grow vegetables for export (notably those in the Valley specialising in artichokes).

Besides potatoes and corn, traditional food crops cultivated also include beans, peas, olluco (Ullucus tuberosus), mashua (*Tropaeolum tuberosum*) and oca (szczawik bulwiasty – *Oxalis tuberosa* L.). The three last tuberous crops have been known since Inca times, and are widely consumed by local people through to the present day. Many other kinds of vegetable are also grown to meet people’s own needs, with fertiliser taking the form of natural manures (Córdova, 2014).

Expressed on a per-farm basis, agricultural output is low. Any potential increases are primarily being limited by periodic water shortages and the land ownership structure. The so-called *minifundio* form of holding is indeed very small, and this stands in the way of the receipt of either technical assistance or credit that might improve efficiency and profitability. Private ownership prevails, so little land is rented. Producers have not organised, and mostly obtain low prices for any of the harvest that they do sell. Viques District has some 491 farms whose total area of farmland is just 138.3 ha. This converts into a mean size of farms equal to 2815.8 m^2, or less than 1/3 ha per family (*IV CENSO NACIONAL AGROPECUARIO* 2012). In reality, just two farms cover more than 4 ha, while as many as 242 are less than 0.5 ha in area. 57% of farms are run by women (allowing menfolk to take up whatever short-term opportunities for work might arise in the city) (*ibid.*).

Traditional cultivation techniques prevail here, though one manifestation of modernisation is the use of artificial fertilisers. In contrast, the tools used out in the fields are in essence the same as those depicted by the Conquistadors. (ZDJęCIE ORKI I RADŁA). Thus farmers will make use of primitive hoes, wooden ploughs, hand ploughs, rakes, yokes, etc. (*yunta*, *picos*, *zapapicos*, *azadón* and *rastrillo* in Spanish). At the busiest times of the year, the whole family works out in the fields. Women play the important role of preparing the fields for ploughing. Water can be in short supply, but irrigation is practised using channels that take advantage of slopes. This means that most of the water that does flow is lost without any use whatever having been made of it. In line with the relief and small sizes of fields, all work is done by hand. Mean output per ha varies greatly, depending on cultivation techniques, soil quality, weather conditions (both precipitation and sunshine) and inputs of labour. Agricultural Census data suggest that the District has lower mean output levels than in the wider region, and most especially when other parts of the same Mantaro Valley are compared.

It is clear that, where the given examples of crop plants grown in Viques District are concerned, only two (grown here traditionally since pre-Columbian times) – i.e., oca and mashua – yield more than in the region as a whole. This attests to ongoing faithfulness to local crops grown using local methods.

Table 2. Mean yields per ha of selected crops

Crop	Yield – t/ha in Viques	Yield – t/ha in the wider Province
Cobs of maize	7.0	12.3
Maize grains (dried)	3.5	2.9
Potatoes	8.5	18.4
Wheat	2.5	2.5
Quinua	2.0	2.7
Oca	6.0	5.5
Mashua	7.0	6.6
Onions	7.0	19.3
Carrots	8.0	23.8

Source: *Dirección de Información Agraria*, Junín, 2014; Maita Franco and Choy González (2004:74).

Any commercialisation of the sale of surplus production is organised on a local basis. Given the small sizes of family farms, these surpluses are of course small, albeit important in the way they supply both farm workers and the inhabitants of local small towns with food items via the weekly markets held. What is more, the great majority of farm owners and their families regard these kinds of sales as a basic source of upkeep.

The District is small and crowded, hence the very limited scale on which livestock rearing is engaged in. There is neither enough pastureland nor enough water to allow for more. The draft animals here are bulls, and there is at least one of these in nearly every household. There are far fewer cows. The *Vista Alegre* sector – above 3200 m a.s.l. – is characterised by the raising of sheep, which locals put out to graze on the sparsely-vegetated summit areas. Llamas are hardly raised here at all, though the cavy (guinea-pig or *cuy*) is ubiquitous, supplying families with its delicate meat. Raising of these rodents is a pre-Columbian practice, and some family farms even specialise in it. There are few head of poultry.

In what is a traditionally agricultural District, industry is lacking; although the Mantaro Valley has an abundance of mineral deposits. Food processing plants operate on a limited scale, e.g., where corn is separated from cobs, or where *chochoca* is produced. Potatoes are dried, wheat is roasted, peas are shelled. The aforementioned *chochoca* (or *chuchoca*) is a form of fine grits made from maize. Cobs are boiled, set out to dry, separated from the grains and then ground up (originally stone-ground). The grits are then added to soups or sauces. In a tradition maintained to this day, colourful cloth is woven here, supplying inhabitants, but also being sold to a limited extent in the course of fairs marking religious holidays. Clay pots and other vessels are also made in a couple of places, as are ornamental ceramics of clay.

The development of tourism is associated with investment pursued by long-term Mayor of Viques Emerson Nolasco, whose mission has been to develop a Water Park here. The development has been favoured by the location in the Mantaro Valley, as well as by the presence of small bodies of water on the lower river terrace. The complex today comprises bathing pools for adults and children, trampolines, slides, baths for men and women, a car park, and an area with eateries serving local dishes (ZDJĘCIE z ośrodka). It is further planned for there to be a sports field and a small botanical garden, as well as further modernisation work done on the existing installations.

The key benefit from the development is the way it has given employment to more than 100 people in a very small locality. Following checks on their incomes and preparedness for work, some 50 women – mainly single mothers – were employed at the Park's eateries. 50 rickshaw drivers represented a further group to benefit from the development, and finally there are also 15 employees who help run the site and maintain it in a clean and tidy state. An exhibition area is now taking shape, as well as a kiosk selling the output of local artisans. Since this kind of complex remains a rarity in Peruvian cities, the Mayupampa recreation area receives visits by inhabitants of the nearby city of Huancayo (with its over 300,000 inhabitants). An attraction of Viques that ought to find its way on to the calendar of tourist events in the *Altiplano* is the *Huaylas Huanca* Festival, which features the local dance called the *huaylash.*

Chapter 14

POLLUTION OF THE NATURAL ENVIRONMENT AT LOCAL LEVEL

As is made clear in the District's development plan (Córdova et al. 2014), the existing potential for development (shaped by natural and cultural conditions) is not just used improperly or poorly, but is also in fact threatened by actions intended or unintended on the part of local people and local authorities. For example, the burning of pastureland in the more elevated parts of the District leads to the destruction of natural vegetation, and thence to more limited retention of water. In turn, over-intensive irrigation techniques in fields lead to soil erosion; while the pollution of the Mantaro by mineral compounds arising from the exploitation of mineral resources in the upper stretches leads to still-wider contamination of both soil and water by heavy metals, and so on.

The location of Viques District in the *Altiplano*, at altitudes in excess of 3000 m a.s.l., ensures that this area typically has a cool montane climate with major variations in weather conditions even in the course of a 24-hour period. Summers are very wet here, often with high rainfall totals (and rarely also snow). Winters in turn tend to be dry. Soils at this altitude are prone to degradation, so their management needs to be engaged in with particular care and skill.

While applying traditional techniques and methods of production, farm management in these areas cannot be said to constitute, or be leading to, sustainability. Resort to a system that entails both rotation and burning, irrigation based solely around flow down slope inclines and extensive grazing of livestock on poor pasture are all leading to further degradation of soil, and damage to existing ecosystems. Such a weakening of natural ecosystems increases vulnerability of the land to extreme weather phenomena, as well as inappropriate agrotechnical techniques and measures.

As in Frias, so also in Viques District, it proved possible to identify a series of threats to local development, albeit ones of differing origin and significance. These are detailed below.

(a) Extremely heavy (non-October-April(May), i.e., non-rainy-season) rainfall associated with the *El Niño* event, and probably made still worse by ongoing climate change. Agriculture is first and foremost put at risk by the intensified precipitation, though the mudslides and landslides generated also destroy roads (while rising waters have also prevented crossings of the Huaycus Valley). In February 2016, heavy rain in the Andes destroyed many earth roads in Piura Province (including many that had been painstakingly re-stabilised a year before). Mudslides lead to the destruction of soil layers on the Andean slopes. The area affected by erosion is expanding.

(b) Ground frosts quite often affect land in the District, especially in the more-elevated parts of Viques, above 3400 m a.s.l. Observed climate change seems to be ensuring the appearance of frost at times of the year when it was not present before, with the result that (at least) parts of crops suffer destruction.

(c) Droughts appear more and more frequently, thanks to delayed onset of the rainy season. They are most typical for October, but sometimes also November. Deforestation of land here has favoured changes in water relations. Where rain is lacking in areas above 3000 m a.s.l., harvests of potatoes (the staple diet of people residing at such altitudes) may be put at risk. Droughts are often accompanied by strong winds, which can destroy houses not made of more-permanent materials. Winds may also blow away soils, sometimes also with newly-sown crops (e.g., of maize or beans).

(d) Pollution of the environment, and especially of flowing waters or lagoons, which may pose a serious threat to human health. Some pollution is also caused by improper waste management in the area. Still more serious, and wider-ranging (whole-region) consequences may arise from pollution of the Mantaro with mine wastes discharged into the river in its upper course (where ore deposits are exploited), and then carried down the Valley before being dispersed on to crop fields.

In districts lacking larger industrial or infrastructural developments, pollution of the environment results first and foremost from the non-functioning or improper functioning of systems for the dumping and utilisation of household wastes.

Chapter 15

Factors Influencing Territorial Sensitivity and Obstructing Sustainable Resource Management

Like Frias before it, Viques was the subject of our fieldwork carried out from December 2013 onwards, with a view to the District's socioeconomic circumstances being diagnosed, and barriers to – or chances for – development identified and defined. Among the key problems to have been diagnosed here is the state of the natural environment. Responses from inhabitants as regards such problems tend to make reference to a lack of effective waste management (40%), the pollution with oil of the soil and waters (16%), the polluting of the River Mantaro (4%), a lack of water in general (6%) and a severe deforestation problem (4%). Other answers concerned the lack of infrastructure and job opportunities, the inadequacy of medical care, a low level of education and so on. In fact, this is rather a repetition of issues met with earlier in Frias, which is to say that reference is made to:

- a lack of appropriate medical care and information on how medical assistance can be received;
- a public transport system that is chaotic and ineffective;
- a continuing lack of installations and infrastructure by which to attract – and keep – weekend tourists (notwithstanding the setting up of the recreation centre);

- a lack of effective waste management;
- the lack of an organisation that would support the commercialisation of local handicrafts (notwithstanding the tradition of such crafts that is present, and the original nature of what is made);
- a lack of appropriate promotion of cultural traditions (including those mentioned in the previous point, but also extending to elements like local folk dances);
- a low level of education;
- a lack of water for households and for the irrigation of fields;
- serious social problems associated with alcoholism (including among those who are under-age), as well as discrimination against migrants.

Furthermore, economic and social problems difficult to resolve at local level obviously exert an impact on the way in which resources are managed, and on relationships between human beings and the natural environment. The table below presents conclusions arising out of the public consultations and interviews run in Viques in 2013 and 2014, and relating mainly to the District's most serious environmental problems.

The following issues were mentioned by those surveyed, among other problems and weaknesses of the District that represent a major obstacle to the improvement of living conditions and raising of people's incomes.

1. The lack of an organisation representing agricultural producers, which are deprived, not only of good administration, but also of technical assistance and support, and capital for their activity. This situation limits the competitiveness of local farmers, as well as
2. The degradation of natural resources – soil and biodiversity, occurring thanks to a lack of suitable regulations, and/or their enforcement; as well as ineffective action on the part of the local administration.

Table 3. Natural resources and environment

Problem	Cause	Factors favouring strategy	Threats
Disappearance of local species of tree and a shortage of wood to meet people's needs	1. Cutting of trees for fuelwood since pre-Columbian times. 2. Lack of reafforestation programmes. 3. Lack of educational activity regarding the significance of forests for the region's natural environment.	1. Existence of land suitable for (re)afforestation. 2. Existence of native species of tree adapted to a cool montane climate.	1. Serious climate change exerting an unfavourable influence on water relations and weather in the region. 2. Fires started by inhabitants burning off grass. 3. Unskilled tending of nurseries and newly-planted trees.
Decline in the supply of medicinal plants and herbs.	Application of pesticides and other chemical agents contributing to the disappearance of many field weeds and herbs.	Persistence of certain herb and medicinal plant species, for which demand is rising on urban markets.	Disappearance of certain plants and the lack of a possibility for them to be reintroduced.
Severe pollution of the environment caused mainly by the spread of sites at which waste matter is accumulated and processed.	1. Dumping of household waste and litter. 2. Poor organisation of waste collection.	A small part of the District should provide for better organisation of waste management.	1. Threat of disease due to the ubiquitous presence of refuse and other kinds of waste. 2. Worsening pollution of the District's flowing and standing waters.
Over-intensive use of natural resources, with simultaneous neglect of, or disdain for, the consequences as regards the natural environment.	Autonomous, uncontrolled felling of trees and prospecting for raw materials, use of water in lagoons for irrigating fields.	Planting of the lowest river terrace with fast-growing eucalyptus trees capable of supplying wood for daily use.	1. Invasion of alien plant species. 2. Lack of popular acceptance for programmes seeking sustainable resource management.

Source: H. Cordova-Aguilar, 2014: 78

3. Agricultural production that lacks added value; the use of low-quality grain and primitive farm implements.
4. The lack of appropriate measures to prevent damage arising in agriculture, as well as to manage crises – on the part of local authorities, state institutions and organisations in society.
5. The poor state of technical infrastructure, which is further damaged and subject to breakdowns during the rainy season.
6. Inadequate levels of provisioning as regards healthcare, education and other social services.

Chapter 16

PROGRAMMES FOR THE MODERNISATION OF AGRICULTURE

One of the main tasks when it came to raising the productivity of Andean agriculture (and hence increasing the income and improving the living conditions of the region's rural inhabitants) – was the introduction (*i.a.* by international organisations including the FAO) of programmes and projects offering technical assistance to countries of the global "South" (Bourlaug 2000: 4). By definition, these were to be programmes increasing efficiency and raising agricultural output per hectare. Up to the end of the 20th century, the results of most of these programmes were a matter of debate. In some countries they did achieve the results anticipated for them (especially in the early phase of implementation of international projects like the "Green Revolution"). However, in a great many states they ended in fiasco, mainly because of a lack of implementation of the environmental projects that were supposed to go hand in hand with the agricultural work. For example, an irrigation project for central Niger led to environmental change and a worsening in the living conditions of the area in question's inhabitants. Modern technologies applied very often in the global "South" excluded – and undermined the sense of – these projects, since they did not take account of local environmental, social and cultural content , most especially by downplaying the role of traditional farming techniques often emerging as the only ones appropriate for application in a given environment.

In the case of Latin America, one of the reasons for the projects and programmes for the development of agriculture implemented in rural areas in the 20th century to be criticised concerns a lack of real impact in raising standards of living, and hence combating poverty. Garcés Jaramillo (2011) claims that a prevailing opinion to the effect that rural poverty reflects issues with agricultural output was responsible for the search for solutions entailing "increased farm productivity" – primarily through an increase in the area under commercial crops requiring large inputs of capital – in place of low-productivity subsistence systems (Garcés Jaramillo 2011: 18). At the same time, from the 1990 onwards there was a dissemination (promotion) of models of rural development and agriculture based on production for export as the only ones capable of helping put an end to poverty. The macroeconomic model adopted laid emphasis on economic growth, without account being taken of the distribution of means or the costs to the natural environment (which in essence gave rise to the situation described by the term "growth with poverty" (*crecimiento con pobreza*) (Altieri 1992: 2).

L. Martínez (2002) claimed that – in the case of communities of indigenous peoples – economic development assumed a form different from that recommended by the World Bank (which introduced development programmes in South America). It was noted that disparities and stratification in communities tended to give way to greater homogeneity, at least in the sense of the use made of income obtained. The process in question is a search for sources of income other than agriculture, with work to earn that income engaging all active members of the family. Different strategies for earning additional income are selected, with these including emigration for the purposes of gainful work, mini-production on the internal market (e.g., of cheeses, fruit preserves, jams, and so on), the sale of craft products, agritourism and others. Rural communities strive to decouple their income from the level of agricultural output, seeking other sources instead. New legal regulations on the rural economy, and especially access to microcredits and other financial instruments, represent one means by which to raise the competitiveness of the areas under study on the domestic market. At the same time, penetration of rural areas by commercial companies and intermediaries ensures that small producers are at risk of being exploited, and simply do not know how they might be protected or defended. On the other hand, they lack the kind of organisational skills that might help with that, and on the other the parties as such are not organised to promote their goods on the domestic and international market. The external market – and first and foremost the state – should support small farmers, producers of traditional food products working to enhance their own internal market for these, and so seeking products whose appropriate promotion might even see them becoming products for export. [An example of good practice is the support offered by the wife of the President of Peru at UN fora, where she presented the health benefits of consuming *quinua*].

Despite retaining traditional agriculture in many parts, Andean rural areas have also experienced unfavourable effects of commercialisation, modernisation and a certain lack of information and training for farmers as regards the consequences of applying modern agrotechnical measures. For example, as a result of the excessive use of fertilisers in a region where research was carried out, a decline in soil quality has ensued. Elsewhere farmers have been using water from the Rio Montego polluted with mine effluents to irrigate their fields; while in yet another place the effect of cutting down trees to make way for crops or to supply fuelwood has been to reduce the level of biodiversity, etc. Unfavourable social processes are to be added to this picture. Garcés Jaramillo (2011: 19) refers to the indirectly or directly unfavourable impact of changes in the environment on the health of farmers, the collapse of certain social relations at local level, the disappearance of traditional practices and skills applied in the Andean villages, and so on. In line with the research carried out by the author cited, the modernisation of agriculture has especially entailed the inclusion of domestic producers and international ones active in South America into the global systems dedicated to the production of soybeans, cotton, flowers, certain vegetables, etc. The effect has been huge damage done to the natural environment. Likewise, the expansion of the area devoted to commercial crop-growing in Ecuador became the main cause of degradation of traditional production systems, as well as a reduction in biodiversity. By the same token, an increase in the area of arable land has given rise to serious environmental problems (PNUMA-FLACSO-MAE 2008: 15-18). The situation looks similar in Bolivia, where eastern forested ecosystems are disappearing under pressure from producers of soybeans and cotton.

In a circumstance of threats being posed to the very basis for their existence, Andean farmers seek other sources of employment or a diversification of their activity in order to have a wider range of assets at their disposal. Farming does not guarantee them appropriate income to remain in the region. Most rural families are forced to widen the range of activities engaged in and the sources of income. This is possible through strong family ties, as well as relationships within the local community whose members help one another. The disparities in ways of life and of earning income thus have both a social and an economic dimension.

The situation today in rural areas of the Andean states reflects the action of a series of factors, among which a leading role is played by inequality and the conflicting interests of different groups of inhabitant present since the colonial era. The complexity of the situation also arises out of different forms of land management, ownership or tenure, with the result that studies of under-development and the different practices seeking to promote development demand precise regional and local analyses. Farming systems also differ to some extent from one Andean country to another, even if there is enough overall similarity to allow for some generalisations in regard to the region as a whole. Thus, for example, in Ecuador in 2001 it was possible to identify several types of production strategy being pursued by farmers engaged in subsistence agriculture, as well as joint or common management of local resources of the environment (Flora 2001). The author in turn worked alongside her team to study the Ecuadorean farmers with a view to gaining a fuller acquaintanceship with their strategies supporting or assisting the sustainable development of agriculture and management of environmental resources, by way of social participation in both processes. A hierarchical system of strategies was identified here, as linked with access to farm land, the land-use structure, the main economic activities of the local community, the structure to crop and livestock production, structure as regards assets and work done outside agriculture. The combination of all these factors allowed the author cited above to identify and distinguish between 7 different types of production strategy adopted by local farmers, who are:

- producers supplying an external market (commercial crops like sugar cane);
- medium-sized producers characterised by a diversified production structure;
- owners of land in remote regions and/or not very suitable for cultivation;
- small-scale producers making frequent changes of direction as regards farming output;
- small-scale breeders of livestock;
- leasers of land not otherwise in possession of it;
- farmworkers with no land (Flora et al., 2001 after Garcés Jaramillo 2011: 33).

It results from the above that there is no consistent breakdown here, but rather one based on markedly variant criteria concerning forms of functioning of the management engaged in in rural areas.

In the case of work on rural communities in the Peruvian Andes, the reference point comprised the three selected localities of Frias, Viques and San Jose de Surco, which differ as regards their forms of management, as well as conditions where their physical geography is concerned. In-depth interviews were run and questionnaires supplied at the level of the household, most of all in Frias, where there were almost 400 cases. In the other localities, a series of in-depth interviews were held with local leaders and those running small businesses (for example in San Jose de Surco with those trading in flowers, with the leaders of the indigenous Amerindian community and with those making yoghurt on small farms). The primary objects of study were agrosystems of the traditional type, though these are rarely present in pure form, and most often also display features of conventional agrosystems, as well as organic farming in some cases.

There is a wealth of literature devoted to organic farming, and above all to the related issue of the sustainable development of rural areas (e.g., Altieri 1999…..). It is most common for work on the latter topic to have been done from a political-sciences point of view (Garcés Jaramillo 2011). In turn, when it comes to the agricultural policies of Andean countries, key problems associated with the expansion of farming relate to the loss of biodiversity in the face of ever-expanding monocultures. A Lack of diversity where cultivation structure is concerned ensures that production systems in place over larger areas are very vulnerable to negative external impacts, especially crop diseases (Altieri 1999).

Indeed, it was the assertion of Altieri (1999) that the examples of rural-development and agricultural programmes in Latin America were making it clear how maintained or even raised biodiversity in traditional agrosystems could be equated with the safeguarding of a more varied diet, as well as diversified sources of income for local people (Altieri 1999: 310). It simultaneously offered a guarantee as regards stabilised levels of agricultural output, with a reduced risk of losses at harvest time; and it favoured an intensive form of production in these particular ecosystems, notwithstanding limited outlays of capital from local small farmers who manage the land here. Traditional cultivation techniques and management in general applied by local producers for generations allow for the maximum yields possible in this kind of agrosystem.

Needing to be added to all this is the cultural and religious conditioning upholding management traditions that follow the changing seasons of the year and respect natural conditions. The consequence is that viability and income on such farms is guaranteed, while natural resources are preserved in a non-deteriorating state (Altieri 1999: 310). Researchers dealing with the efficiency of traditional forms of management stress that – with "modern" agriculture, the farmer receives a package of instructions and guidelines on how to run a farm; while within the wider agroecological system that farmer's procedures are based on the experience gained over many years of management in practice, often passed down from generation to generation. For Altieri, agroecology represents a set of management principles that pay heed to and respect local natural and cultural conditions (cited after Amorin 2008).

In her studies of traditional agriculture and its role in preserving local identity and traditional farming culture, Susan Hecht (1999) mentions three main causal factors that have pushed traditional systems of cultivation to the margins of agrarian life in the countries of Latin America. These are:

1. the breakdown of principles or rules in line with which farming practices are codified, regulated and passed on;
2. dramatic changes ongoing in many Amerindian communities and in the production systems upon which their existence was once based – these changes *inter alia* reflecting demographic breakdown, slavery and colonialism, as well as market processes;
3. the predominance of a positivist approach in discourse on the subject of development.

In this situation there were few opportunities to ensure that holistic thinking and behaviour based around intuition developed in traditional agricultural systems would gain acceptance among technocrats whose goal was to maximise output at all costs (Hecht 1999: 15). In the period in which many "Third World" countries were introducing programmes arising out of assumptions regarding the "Green Revolution", there was a prevailing view that practices small farmers engaged in might improve the situation on farms in the long term, while the agricultural modernisation programmes assumed the achievement of clear growth effects within just a short time (Gliessman et al., 1981 as cited by Altieri 1999: 103).

Part II. Biodiversity Management as an Adaptation Strategy to Climate Change

Chapter 17

BIODIVERSITY OF THE CENTRAL ANDES

The Central Andes stretch the section of the cordillera that crosses Peru longitudinally; then, to talk of the Central Andes is to refer to the Peruvian sierra. This sierra stretches from the 3°30'S and the 18°21'S. Its ecosystems are strongly influenced by the Pacific light winds that blow to the north and by the trade winds that blow from the Atlantic, recharging moisture as they pass the La Plata and Amazon rainforest to discharge at the east side of the Andes. These global conditions favor the existence of a great ecosystems diversity (Villegas Nava, 2012) which holds the great biodiversity reported by scientists who have studied this topic in Perú (Brack Egg and Mendiola Vargas, 2000). According to Brack Egg and Mendiola Vargas (2000:390-1) in Peru there have been identified near 25,000 native plant species, of which 17,143 are angiosperms. There are 128 domesticated native plant species, some of them with thousands of varieties, such as the potato, which at present is one the most important crops of the world, together with rice, corn and wheat. There are now 6,034 native plant species used in Peru and from them there are 710 food species and 1,109 medicinal species. Not all are Andean because here are also included the ones from the Amazon. However, it is important to note that there is a great number that not only are accepted and consumed by the rural households, but because of their nutritional conditions and flavor they have an enormous potential to control de urban market. Veerle van den

Eynden, Cueva and Cabrera (1998) report that in southern Ecuador they have identified 250 wild species that are being used as food by the native population. Many of these species are also found in northwest Peru with similar uses. Tapia (2000: 21-3) includes a list of edible Andean plants which are shown in the table below.

Table 4. Native food plant species of the Andean region

Common Name	Scientific Name	Family	Altitude of Optimal Growth (m.a.s.l)
***Tubers**:*			
Potato	*Solanum andígenum*	Solanaceae	1000 - 3900
Bitter Potato	*Solanum juzepczukii*	Solanaceae	3900 - 4200
Oca	*Oxalis tuberosa*	Oxalidaceae	1000 - 4000
Olluco, Ulluco, Papalisa	*Ullucus tuberosus*	Baselaceae	1000 - 4000
Mashua, Isaño, Año	*Tropaeolum tuberosum*	Tropaeolaceae	1000 - 4000
Roots:			
Arracacha, White Carrot	*Arracacia xanthorhyza*	Umbeliferae	1000 - 3000
Achira	*Canna edulis*	Cannaceae	1000 - 2500
Jícama, Ajipa	*Pachyrhizus tuberosus*	Fabaceae	1000 - 2000
Yacón, Aricoma, Jiquima	*Esmalanthus sochifolia*	Compuestae	1000 - 2500
Chago, Mauka, Miso	*Mirabilis expansa*	Nyctaginaceae	1000 - 2500
Sweet Potato, Apichu	*Ipomoea batata*	Comvolvulacea	0 - 2800
Maca	*Lepidium meyenii*	Cruciferaceae	3900 -4100
Grains:			
Mays, Sara	*Zea mays*	Gramineae	0 - 3000
Quinoa	*Chenopodium quinoa*	Chenopodiaceae	0 - 3900
Qañiwa	*Chenopodium pallidicaule*	Chenopodiaceae	3200 - 4100

Common Name	Scientific Name	Family	Altitude of Optimal Growth (m.a.s.l)
Amarant, Coyo, Kiwicha	*Amaranthus caudatus*	Amarantaceae	0 - 3000
Fabaceae:			
Tarwi, Chocho	*Lupinus mutabilis*	Fabaceae	500 - 3800
Bean, Poroto	*Phaseolus vulgaris*	Fabaceae	100 - 500
Pallar	*Phaseolus lunatus*	Fabaceae	0 - 2500
Pajuro	*Erythrina edulis*	Fabaceae	500 - 2700
Cucurbitáceae:			
Squash	*Cucurbita máxima*	Cucurbitaceae	500 - 2800
Caigua, Achoqcha	*Ciclanthera pedata*	Cucurbitaceae	100 - 2500
Fruit:			
Pepino, Kachum	*Solanum variegatum*	Solanaceae	500 - 2500
Mataserrano	*Solanum muricatum*	Solanaceae	0 - 2500
Capulí, Uchuva, Uvilla	*Physalis peruviana*	Solanaceae	500 - 2500
Sachatomate, tree tomato	*Cyphomandra betacea*	Solanaceae	1800 - 3000
Granadilla	*Passiflora ligularis*	Passifloraceae	800 - 3300
Tumbo, Curuba, Tumbito	*Passiflora mollisima*	Passifloraceae	2000 - 3000
Tumbo, Granadilla Real	*Passiflora quadrangularis*	Passifloraceae	0 - 2500
Chirimoya	*Annona cherimola*	Annonaceae	1000 - 3000
Lúcuma	*Pouteria obovata*	Sapotaceae	0 - 2500
Pasakana, Ulala	*Eriocereus tephracentus*	Cactaceae	500 - 2500
Pasakana de Chuquisaca	*Trichocereus herzoqianus*	Cactaceae	500 - 2500
Chicope, Papayita arequipeña	*Carica pubescens*	Caricaceae	1000 - 3000
Mora de Castilla	*Rubus glaucus*	Rosaceae	2000 - 3000
Ciruela del fraile	*Bunchsia armeniaca*	Malpigiaceae	500 - 2500
Sanky, Sancayo	*Corryocactus brevisylus*	Cactaceae	2500 - 3200

Source: Tapia, 2000:21-23; and author's additions.

This plant biodiversity is complimented by the animal biodiversity which shape the Andean ecosystems. Both they are an important source of products to the sustainment of the local human population (fish, animal game, fruit, medicinal plants, fibers, handicrafts, firewood, ink, dyes, wood, tourism, and others).

Literature related to useful plants to mankind originated in the Andes is extensive, especially related to food production (León, 1964; Cárdenas, 1969; National Research Council, 1989; Ayala, 1992; Hernández and León, 1992; Tapia, 2000). In regard to the Peruvian sector we may mention the work of Mario Tapia whose research goes back from 1960 to present. In the 1980's the FAO invited the governments to dedicate more interest to native crops, and do more research to determine their economic and nutritional importance. This was because the agricultural advisors found that the Andean peasants dedicate too much labor in their native subsistence crops (Morón, in Tapia, 2000:3). But the interest for Andean cultigens goes back to the first quarter of XXth Century and was lead by botanists and agronomists. However, the major attention was placed in 1980 and following years, due especially to the insufficient food production based on the traditional food sources and the search for answers to solve hunger and undernutrition of the world population, and in particular the Andean, where up to 2010 there were around 30 million poor and under-nutritious in Latin America (Tapia, 2000; Tapia and fries, 2007:1).

Here, we are more interested in two topics: a) to show the nutritional potential of some native wild Andean fruit that are being used by the rurals in their household diets, but without any intention to go into their agriculture; b) to mention the medical potential of some of these Andean species that are being used by the rural local population. Some fruit have already called on the rural attention to incorporate them in agriculture, such as the cherimoya (*Annona cherimolia*), guayaba (*Psydium guayava*), aguaymanto/uchuva or capulí (*Physalis peruviana*), capulí (*Prunus serotina*) tuna (*Opuntia tuna*), pacae (*Inga feuillei*), lúcuma (*Pouteria obovata*), pepino dulce *(Solanum muricatum* Ait.), tomate de árbol *(Cyphomandra betacea* Cav. Send) and others. There are still many others which may be incorporated in the people's diets with the consequent monetary additions to the household's income. Here are the pitaya (*Hylocereus ocamponis*), zarzamora (*Rubus robustus*), lulo (*Solanum quitoense*), guaba (*Inga* (two varieties)[1], arrayán (*Eugenia quebradensis*), masaugache (n.i), tumbo (two varieties: *Passiflora mollisima* and granadilla real (P. *quadrangularis*), chicope or chamburu (*Carica pubescens Linneo & Koch*), toronche (*Carica stipulata Badillo*), sancayo or sanky (*Corryocactus brevistylus)*, mito (*Carica candicans*), ciruela del fraile (*Bunchosia sp*), guayabillo (*Psidium friedrichsthalium*), chamelico (n.i), etcetera.

[1] There are near 200 species of the Inga genera in Souh America (Francis, 1994).

Chapter 18

UNDER USED FRUIT IN THE CENTRAL ANDES

As mentioned above, there is a good number of vegetal species that give fruit with vitamins and other nutritious sources to the rural population, which at great extent are unknown by the urban householders. Not all they have been botanically identified already even though their nutritional values have been recognized by the local people at places where they grow.

Arrayán (*Eugenia quebradensis*, *Eugenia sp.*). It is a tall shrub of 4 to 6 m high, much branched with dark green leaves; the upper side of leaf is bright and the down side is opaque. Flowers are white, and the fruit rounded of 1- 2 cm is green when young and change to black when ripe. It grows in humid environments and is part of the foggy forest of the high Yunga and low Quechua between 1500 to 2800 m altitude. Eynden, Cueva and Cabrera (1998: 106) report to have seen this tree in the western side of southern Ecuador. It is used as fresh fruit and has a resemblance to grape but its skin is rougher. It has a good potential to place it in the market as jam, refreshment and ice cream.

Figure 1. Fruit of arrayán. (Source: Internet).

Chicope (*Carica pubescens, Linne & Koch).* Also known as "papayita de olor", "papaya arequipeña", "chamburu" (Ecuador), "papaya del monte", "papaya de altura", "papayuela" (Colombia), is a caricaceae native to the Andes which is actually cultivated in household gardens from Colombia to Bolivia. The history of this Andean fruit is not well known but probably it was extracted from the perennial Andean forest; it also seems to have been cultivated before the introduction of *C. papaya* (Sánchez, 1992; Tapia, 2000:110).

Figure 2. Fruit of *Carica pubescens*. (Source: H. Córdova).

Chicope is a species of erect stem, with few branches of some 8 m height. It grows in humid zones, near the water courses between 1500 and 3000 m altitude. The altitudinal level depends on latitude, the lower altitudinal levels are related to lower latitude levels. It enjoys annual rainfall ranges from 500 to 1000 mm and media temperatures of 12 to 18°C. This species is sensible to low temperatures and to the intensive sunshine at noon; it does not resist long drought periods. Fruit is a big berry –as seen in Figure 7- oblong like the Hawaiian papaya of green skin, which turns yellow when ripe. It is mostly consumed at household level as fresh fruit in refreshments and jam (van den Eynden, Cueva and Cabrera, 1998:90). When this fruit green is also cooked and eaten as vegetable. Besides these consumption modes, this fruit is also used in industry to extract its latex that contains papain applied to tender meat and against the skin mycosis and plain wart. In folk medicine, the fruit is good against arteriosclerosis, diabetes and hepatic illnesses. The nutritional composition is shown in the following table:

Table 5. Nutritional composition of chicope (100 g of pulp)

Compound	*Amount*
Energyz	24 Kcal
Water	93.0 %
Protein	0.8 g
Fat	0.2 g
Carbohydrates	5.4 g
Fiber	0.0 g
Ash	0.6 g
Calcium	12.0 mg
Phosphorus	14.0 mg
Iron	0.5 mg
Thiamin	0.02 mg
Riboflavin	0.03 mg
Niacin	1.0 mg
Ascorbic Acid	47.0 mg

Source: INCAP. *Tabla de composición de alimentos de Centroamérica*, 2012:43

The chicope is already cultivated at commercial scale in Colombia, Chile, and northern Ecuador; in Peru it is stillhold as a marginal crop that grows almost wild and then its supply to urban markets is small. It is found along the Quechua region at both sides of the Central Andes.

La Ciruela del Fraile (*Bunchosia armeniaca, Cav*). This is a tree of some 8 meters high with woody and hard stem. Its fruit is an oval drupe, fleshy that when ripe it is red and edible. This plant is cultivated in Perú since pre-Hispanic times, especially at the Yungas from 500 to 2500 m altitude, but only to local household consumption. The Spanish conquerors found this species in those environments and Oviedo (cited by Yacovleff and Herrera, 1933:27) called on the attention that "there are certain trees that the Spaniards call plums of two stones -ciruelas de dos cuescos- ; which are big trees and its fruit is properly as plum, and each of them has two stones; and the Indian eat them as well as the Christians, though its flavor is less than good and the fruit flesh gets sticky in teeth". Maybe this is why the fruit is also called "cansaboca" in some parts of Peru. In Colombia is also known as "uvito morado" (Jiménez-Escobar, et al, n.d:401) and consumed as fruit.

Figure 3. Fruit of *Bunchosia armeniaca* (Cav) DC. (Source: H. Córdova).

There was not possible to find any data on the biochemical composition of *Bunchosia armeniaca*, but it is one of the wood species to take into account when planning reforestation programs in the Andean side-slopes to confront climate change.

Guaba de Zorro or Shimbillo(*Inga insignis*, *Inga sp*). It is small tree of some 8 m tall with the stem sometimes twisted and trunk of 2 m long and the frondose umbrella-shaped tree top makes this species much appreciated to provide shadow to the coffee plantations. Its leaves are light brown and the fruit is cylindrical of 10 to 15 cm long which get stuffed when ripe with some 2.5 cm diameter. Inside the husk are the black-light green seeds covered by a white sweet pulp. Only the sweet pulp is edible. This tree is found wild between 800 and 2300 m altitude and it grows well in silt loam soils with good drainage, moisture and temperatures ranging from 15 to 22°C. In southern Ecuador, this tree is found between 100 and 2000 m altitude and its fruit reach 10 to 25 cm long (Van den Eynden, Cueva and Cabrera, 1998:48).

The guaba wood is used as firewood, construction material and to nitrogen fixation in soil due to its ability to hostage the bacteria *rhyzobium* that traps the air nitrogen and makes it available to plant roots. In this sense, this tree is also potentially important to include in reforestation programs.

Another guaba variety that also grows wild between the 200 and 1200 m altitude is the *Inga silanchensis*. This tree may reach 15 m high and produces green color, glabrous, flattened, fruit of 25 cm long and 3 cm wide (van den Eynden, Cueva and Cabrera, 1998: 48), with white flesh that covers the blackish seeds. It is found en abandoned agricultural fields and it is also the fruit tree of household gardens.

The habitat conditions of this species as well as the *Inga insignis*, allow using them in reforestation of the Andean side slopes of the Yungas at both the West and Eastern sides.

Mito (*Carica candicans*, *Vasconcellea candicans*, *Gray A. Gray*). Also known as "mashuque", "jerju", "papaya silvestre", and "quemish" is a small tree of less than 6 m height, deciduous, native to the arid western, inter Andean and amazon yungas of the Central Andes at altitude ranging from 500 to 2000 meters. The fruit is a berry hanging from a pedicel 5-7 cm long, ovoid-elongated, attenuated at the base and the apex, 10 to 20 cm long and 3 to 5 cm diameter, somehow rough skin, green leaden that changes to yellowish when it is ripe and exudes a fragrant smell. It contains numerous seeds, ovoid of some 8 mm long without sarcotesta (viscose external shell), the inside layer brown and almost smooth when dry. It is consumed fresh, after eliminating the latex which is done by putting the entire fruit on the kitchen fireboard to burn the external skin layer. It also is consumed in jam, refreshments, and pastries. The fruit juice as well as leaves is used to soften hard meat by boiling them together. In folk medicine, the ripe fruit eases digestion and the latex is applied to eliminate warts and ulcers produced by leishmaniasis (Soukup, 1970).

According to Dori Felles (2013:9) the fruit has high amounts of total proteins (8,2% in dry weight) and carbon hydrates (70,1%) and the considerable amounts of vitamin C (45 mg) and minerals (Fe, Ca, P, Na, Mg and chlorides). The following table shows the nutritional values.

Table 6. Nutritional composition of fresh mito (100 g of pulp)

Compound	*Amount in %*
Water	88.8
Protein	0.9
Fat	0.3
Carbohydrates	7.9
Fiber	1.2
Ash	0.9
Calcium	15.0 mg
Phosphorus	13.1 mg
Iron	0.5 mg
Ascorbic Acid	45.0 mg
Magnesium	10.0
Sodium	2.9
Chloride	80.0

Sourcee: De Feo, et al. 1999:3683.

Figure 4. Fruit and plant of mito. (Source: Internet and H. Córdova).

The mito seeds contain an oil rich in fat acids, especially those called caprylic, lauric, palmitic, stearic, oleic and linolenic; which suggest a possible use in human nutrition.

The mito was already known and consumed as fruit since pre-Hispanic times (Sagástegui, Rodríguez & Arroyo, 2007) but its consumption has been reduced to the rural household level. It is not cultivated and in Peru it grows wild, and thus it is suffering a depopulation process due to cattle overgrazing and its use as firewood though it is a poor quality wood. The phytogenetic qualities of the Caricaceae are calling on the attention to the scientific community and there are already some risk alarms to its disappearance from some Peruvian places. Its repercussions may be not only in the biodiversity conservation but also in rural nutrition because this fruit is source of vitamins to the rural, especially children (Sagástegui, Rodríguez and Arroyo, 2007).

Mote Mote (*Allophylus mollis Kunth*). This is a tree 8 m high, branchy, with inflorescence cane-tomentose and small flowers grouped in panicles generally terminal. Its fruit is a nut-like indehiscent ovoid-globular, erect, with a fleshy aril somewhat yellowish as well as the flesh which is cremay. After harvest this fruit tends to change color toward the reddish (Zapata Cruz, 2007). It grows semi-cultivated at borders of agricultural fields and wild in the rocky and semi-xerophyte soils along the water streams' terraces and depressions (Sagástegui, 1995; Zapata Cruz, 2007) between 1500 and 2500 m a.s.l, in the departments of La Libertad and Cajamarca, north Peru. It grows quite well in places with media temperatures of 20 to 22°C. Its fruits are consumed fresh by the local householders and they are not sold in the urban markets.

The nutrition qualities of mote-mote (Table 4) put it in the list of potential species for reforestation programs along the sideslopes of the upper Yunga and some sectors of the Quechua region.

Naranjilla or Lulo (*Solanum quitoense*). It is a semi-wild plant species that grows in the East side of the Andes, in tropical rainforest environment between 600 and 1500 m altitude, with median temperatures of 20°C to 25°C. It grows well in shaded environments, with well drained soils and good moisture. The stem is robust of around three meter high, woody like, cylinder and hairy with thorns distributed all over even on the leaves. It grows erect and branchy with four to six stems from the soil level. Leaves are alternate, oblong, oval with the upperside dark green and the underside light green with pronounced purple veins. Flowers are white with purple chalice and a yellow button in the center.

The fruit is a balloon-like berry, 4 to 8 cm diameter and weight of 80 to 100g. It is covered by short and dense hair-like prickles that turn off as the fruit ripes. The shell is yellow intense like orange and the flesh is green-yellowish of sauer-sweet flavor with numerous small seeds, like the ones of tomato.

Figure 5. A flowering branch and fruit raceme of mote-mote. (Source: Zapata Cruz, 2007).

Table 7. Nutritional composition of mote-mote (100 g of pulp)

Compound	*Amount*
Energy	74.12 Kcal
Water	78.78
Protein	0.06
Fat	0.57
Carbohydrates	18.31
Fiber	0.43
Ash	1.61
Ascorbic Acid	trazas

Source: Zapata Cruz, 2007.

Figure 6. Flower and naranjilla plant (http:// es.wikipedia.org/wiki/ Solanum_ quitoense and H. Córdova).

The naranjilla is a fruit already cultivated with much success in the northern Andes, including Ecuador; but it is not in Peru where we still have not given importance to it and we can rarely get it in the urban markets. It is eaten as fresh fruit and in juice, fruit concentrates, ice cream, jam, sauces, pastries, etc. It is commercially cultivated in Ecuador, Colombia and Venezuela and it is exported to the North America and European markets (http://www.fao.org/inpho_archive/content/documents/vlibrary/ae620s/pfrescos/lulo.htm).

This fruit is rich in vitamin C and also has calcium and phosphorous among its principal nutrients, and thus it is said that it contains medical properties that help the good functioning of kidneys. It is recommended to persons who suffer from gout because it helps the elimination of uric acid from the blood. Its nutritional values are given in the following table.

Table 8. Nutritional composition of Naranjilla O Lulo (100 g of pulpa)

Compound	Amount	
	Lulo of Castilla	*Lulo of Selva*
Water	87.0 %	88.0 %
Protein	0.74 %	0.68 %
Fat	0.17 %	0.16 %
Ashes	0.95 %	0.82 %
Carbohydrates	8.0 %	8.0 %
Fiber	2.6 %	2.6 %
Calcium	34.2 mg	48.3 mg
Iron	1.19 mg	0.87 mg
Phosphorous	13.5 mg	25.11 mg
Vitamin C	29.4 mg	30.8 mg
Vitamin E	0.75 mg	0.75 mg

Source: Franco, Germán and others (2002:10).

Pitaya (*Hylocereus ocamponis; H. polyrhizus)*). It is an epiphyte species, green-triangled and elongated stalks; it grows in the branch axis of trees or on the rocks at the deciduos forest of northwestern Perú and South Ecuador (Van den Eynden, Cueva and Cabrera, 1998:38). It also grows in the tropical jungle of the Eastern side of the Andes, far south to the Urubamba Valley, north of the Cuzco department. Its roots are adventitious and grow along the trunk of the hostage tree and at some cases these roots reach the soil. When they grow on rocks their roots extend looking for a soil support to get their nutrients. Flowers are colorful plume shaped of 15 to 20 cm long; light yellow or white with the filaments greenish. The fruit is a berry ovoid green when it is growing and pink when it is ripe; the berry has many bractea foliaceous like the artichoke, but no prickles; when it is ripe it may have 10 – 12 cm long, oblong shaped. Inside there is the pinky flesh with very small black seeds that may be eaten without any difficulty. Of light sweet flavor, much likes the tuna cacti. There are varieties distributed in America from Southern México to the North of Argentina; some varieties have a yellow skin and white flesh, others have a cherry color flesh; but all of them need tropical temperatures. Their reproduction is by stalk or cutting.

(a) (b)

Figure 7. Pitaya plants: (a): pitaya above a rock and (b) wild pitaya fruit (Source: H. Córdova).

The pitaya fruit is consumed fresh and it is a good stomach refresher. It is also prepared on sweets, juice, and jelly, cocktail and even alcoholic beverages in Nicaragua (Muñoz Fonseca, 1997). As it was said, its flesh is eaten with seeds that contain oil that avoids colics and therefore it helps to the good functioning of the stomach and the intestines. The flesh has a substance called captine which helps the good working of heart and as nervous relaxant. The husk is used for animal forrage.

Table 9. Nutritional composition of Pitaya (100 g of pulpa)

Compound	*Amount*
Energy	54
Water	89.40 g
Protein	1.40 g
Total Fat	0.40 g
Carbohydrats	13.20 g
Ashes	0.60 g
Calcium	10mg
Phosphorus	26 mg
Iron	1.30 mg
Tiamin	0.04 mg
Riboflavin	0.04 mg
Niacin	0.30 mg
Ascorbic Acid	8 mg

Source: INCAP 2012: 53; Vite, 2014:37.

This plant is found in tropical environments where the median temperatures are above 20°C; it grows better when temperatures are around the 26°C, if the temperature continues rising the plant production goes down up to 35°C when no longer produces. In Peru, the pitaya grows wild in the deciduous forest of the northwest between the 300 and 800 m altitude. As any other epiphyte it has characteristics that make it resistant to drought and survives to long periods without rainfall or irrigation. The plant needs good natural light to start production. Due to this condition, it is successfully cultivated in southern Mexico, in the Central American countries, Colombia, and Ecuador. It has also been carried to Myanmar, Thailand, Vietnam and other Asian countries where it is commercially cultivated under the name "dragon fruit".

Pushgay (*Vaccinium floribundum, HBK*). Also known as "uva del monte", "uva del campo" or "uvitas"; is a low shrub, rather rampant that reaches about two meters high, with abundant branches, small and rounded leaves. Flowers are white and turn purple when ready to fructification; the roots may reach one meter long and grow horizontally. It is a perennial and starts flowering at the beginning of the rainy season. Its fruit is a small berry, rounded, and comes in raceme; it is blue-violet and its juice is purple colored. It is found wild in the northern Andes of Peru, especially in Cajamarca, though we need more studies to confirm its geographic distribution. Its ecological environment is at the side-slopes of the upper Quechua and Jalca regions which in northern Peru go from 2350 to 3500 m.a.s.l (Tapia & Fries, 2007:119). However, there is a variety known in Huánuco as "gongapa" (*Vaccinium sp)* whose description is similar to the pushgay, and grows up to 4000 m.a.s.l.

This species is drought resistant and to temperatures that may be as low as 10°C. According to Tapia and Fries (op. cit:120) there are four edible varieties: black, reddish, small, and white. These varieties are not only external but also in the flesh color and size. But, be careful, there is toxic variety known as "mio mio" that should be avoided.

Figure 8. Fruit of pushgay (Tapia y Fries, 2007:121).

Sanky (*Corryocatus brachypetalus*). Also known as "sancayo" or "guacaya" is a cactus of 2-4 m high, much branched at the base. The stem is completely wrapped by thorns of 2 to 16 cm long (Britton & Rose, 1914, vol 2: 66); the flowers are yellow cone-like of 4 to 6 cm diameter at its widest section; the fruit is a berry greenish, globular of 6-7 cm of dimeter, wrapped by prickles that get loose and fall when the fruit is ripe and change color to green-yellowish.

This cactus is native to southern Peru, and north of Chile at Parinacota and Tarapacá. In Peru, it is found especially in the Quechua region of the west side of the Arequipa department between 2800 and 3300 m altitude, even though Britton and Rose (1914: vol 2:67) report to have found one individual at 600 m.a.s.l. It grows on the side slopes of dry mountains, at stone, sandy, and rocky places, with scarce water and moisture. It does not resist temperatures above 20°C.

The nutritional value of sanky is shown in the table below, where potassium, calcium and phosphorus are the most significant.

The sanky fruit is sauer-sweet and it is consumed in refreshments, juice, jam, hot beverages, etc. It has remained almost unknown until recently; its name appeared in markets of Arequipa some 10 years ago. It has been incorporated in the gourmet restaurants because it tastes good in a drink called "sankysauer", very similar to "pisco sauer". The concentrated juice is used as laxant and helps to regulate blood pressure, prevents gastritis and liber illnesses. Other applications are against dandruff and to favor hair growth; this is done by washing the head with the sanky husk.

Table 10. Nutritional composition of Sanky (100 g of pulp and husk).

Compound	*Amount*	
	Pulp	*Husk*
Water	95,2%	91,6 %
Protein	1,3 %	1,4%
Fat	0,0 %	0,0 %
Carbohydrates	3,1 %	5,6
Fiber	0,9 %	1,7
Ash	0,4 %	1,4
Calcium	105,5 ppm	752 ppm
Phosphorus	128 ppm	67 ppm
Ascorbic Acid	57,1 mg	2,5 mg
Potasium	5566,4 ppm	1743,9 ppm
Magnesium	145 ppm	
Antioxidant capacity	474,8 ug de eq Trolox/g	

Source: Nolasco & Guevara 2009.

Figure 9. Fruit of Sanky (Source: http://www.perunatural.pe/cabello.html and H. Córdova).

Sauco (*Sambucus peruviana, HBK*). Also known as "lambrán", "layán", and "arrayán" is usually a tall shrub of 3 to 6 m high, but under special environmental conditions it may reach up to 12 m high; perennial.

Its young sprouts are delicate due to their spongy medulla which becomes harder as the stem gets older. The wood is hard and much appreciated as firewood and building material. Leaves are light green, palmate of five to nine folios with sawing like borders; flowers are white racemes and the fruit is a small berry of 5 to 7 mm diameter, green when growing and black when ripe. They are juicy, nice smell and sauer-sweet flavor (Tapia y Fries, 2007: 121). This species is found in humid places at the Andean sideslopes between 2800 and 3800 m of altitude; and its geographical distribution goes from Ecuador to north of Argentina. In Peru it is found along the stretch from Cajamarca to Apurimac, close to the house dwellers and never wild. Many times this shrub is planted aside the acequias or ditches to benefit from the soil moisture there. It grows well in any soil type, though it prefers deep soils, well drained, silt loam with neutral pH or lightly alkali. Once it starts production it continues for various decades. The fruit is consumed fresh and in jams, drinks, pastries, etcetera, with high acceptance by the urban dwellers, and thus it is already sold in urban markets.

The flowers of *S. peruviana* have butyrous essence formed by a terpene and a resin. The bark contains the sambucina alkaloid joined by a resin. The nutritional value of the fruit is showed in the following table.

Figure 10. A species of sauco. (Source: H. Córdova).

Table 11. Nutritional composition of Sauco (100 g of pulp)

Compound	*Amount*
Water	91,49
Protein	1,51
Fat	0,26
Carbohydrates	1,72
Fiber	0,84
Ash	30,6
Calcium	23mg
Phosphorus	1,9mg
Iron	1,6mg
Ascorbic Acid	17,83mg
Vitamin B1	0,07mg
Vitamin B2	0,06mg

Source: Liceth Rocío Huamán Leandro, 2013:10 http:// www.deperu.com/ abc/frutas/5274/el-sauco.

This species is also a good candidate in reforestation programs due to its great ecological adaptation to different soil conditions and because it resists frost quite well.

Tomate de Árbol (*Cyphomandra betacea, Cav.*). Also known as "berenjena", "sachatomate", "yuncatomate", "limatomate", and "tomate de monte" is a Solanaceae native to the Andes where it was domesticated in pre-Hispanic times. In spite of this ancient cultural conquest this species has received little attention and its market area is rather around the production zones. It is cultivated at subsistence level to satisfy the needs of the household family members. It grows well in the upper Yungas and low quechua regions at both sides of the Andes between 1800 and 3200 m a.s.l, with media temperatures of 18 to 22°C and annual precipitations of 600 to 800 mm (Tapia y Fries, 2007: 112). This species has a woody stem erect that may reach 3 m high. Its fruit is a berry oblong of some 8 cm long and 5 cm diameter, greenish or gray that changes to orange red or purple when ripe. It is consumed as vegetable or cooked; in both cases it is necessary to drop off the husk because it is bitter. Other consumption ways is in jams, syrup, juice, sauces mixed with rocoto -hot pepper- (*Capsicum pubescens* R&P), and it replaces tomato in the preparation of stews.

Besides, this species has folk medical applications to soothe respiratory illnesses and to fight anemia due to its nutritional qualities that are shown below:

Figure 11. A tomate de árbol species with *Zea mays* and calabaza (*Cucurbita ficifolia*) (Source: H. Córdova).

The tomate de árbol is found from México to Chile and in Ecuador and Colombia is well known and commercially cultivated. It is exported to the United States, Spain, Germany, Holland and others (Lucas, Maggi and Yagual, 2010-2011:29).

Table 12. Nutritional composition of Tomate De Árbol (100 g of pulp)

Compound	*Amount*
Energy	50 Kcal
Water	85.9 %
Protein	2.2 g
Fat	0.9 g
Carbohydrates	10.3 g
Fiber	0.0 g
Ash	0.7 g

Table 12. (Continued)

Compound	*Amount*
Calcium	9.0 mg
Phosphorus	48.0 mg
Iron	0.8 mg
Thiamin	0.1mg
Riboflavin	0.04 mg
Niacin	1.2 mg
Ascorbic Acid	29.0 mg
Retinol (vitamin A)	300 mg

Source: INCAP. *Tabla de composición de alimentos de Centroamérica*, 2012:38.

Toronche (*Carica stipulata Badillo*). It is a caricaceae species that may reach 10 m high; its stem is rough with thorns, somewhat woody and branchy. Flowers are white, yellowish or redish; fruit is ovoid-oblong, greenish yellowish when ripe; the pulp is creamy yellowish of sauer-sweet flavor, aromatic. The pulp is used to prepare jam and sweeties. It is very rich in papain. It grows at both sides of the Central Andes, especially in the northern sector, that is, south of Ecuador and north Peru between 2000 and 2800 m altitude (Eydnen, Cueva and Cárdenas, 1998:90).

Toronche is an important source of nutrients such as vitamin E and magnesium. It raises significantly the serum levels of glucose, total proteins and globulin. It also diminishes the cholesterol levels significantly with dose of 300 and 60 mg/kg of weight (Muñoz, et al., 2005; Aguirre y Castillo, 2009:6).

Figure 12. Toronche fruit (Aguirre and Castillo, 2009).

Table 13. Nutritional composition of Toronche (100g of pulp)

Compound	*Amaount*
Energy	25.06 Kcal
Water	93.25 g
Protein	0.63 g
Fat	0.22 g
Carbohydrates	5.14 g
Fiber	0.95 g
Ash	0.87 g
Calcium	0.0 mg
Phosphorus	s.i
Iron	2.65 mg
Thiamin	n.i
Riboflavin	n.i
Niacin	n.i
Ascorbic Acid	45 mg
Magnesium	217.22 mg
Vitamin E	45 mg

n.i = no information
Source: Muñoz, et al. 2005:3.

All these studies point out that toronche has great qualities which if used by industry may help to put into value this species and enhance its agricultural interest.

Tumbo (*Passiflora quadrangularis*). It is a vine that climbs on trees or any other support to produce its fruit; which is a big green yellowish berry of 20 to 30 cm long and 12 to 18 cm diameter. The skin is very delicate and the white pulp is light sweet and may be eaten as fresh fruit or in jam, refreshment, sweets, etc. The flesh is smooth and moist and it keeps inside several seeds covered by a watered material, some-like jelly and acidic orange color.

In Peru, it is cultivated from 100 to 2000 m altitude, but it is used only as fresh fruit for household consumption at subsistence level. There are not agricultural programs to introduce this fruit to the urban markets though its flavor is very much like melon.

Tumbo is a tropical plant and it is already cultivated in several countries of Latin America at places with temperatures ranging from 17 to

25°C. It is known under different names such as "granadilla grande", "sandía de pasión", "giant tumbo" and others. In Colombia it is well known as "badea" and it is extensively cultivated as it is shown in next photo.

In health terms, this tumbo contains serotonin which contributes to the good functioning of the nervous system smoothing problems such as anxiety, insomnia, and headache. It also helps to keep bad cholesterol low when eaten regularly. The root is used for eliminating intestinal worms. Leaves are good against arthritis and contusions, knocks and anti-inflammatories; this may be due to the hemolytic activity of saponin.

Figure 13. Tumbo or badea cultivated in Colombia. (Source: Internet and H. Córdova).

Table 14. Nutritional composition of Tumbo (100 g of pulp)

Compound	*Amount*
Water	94.4 g
Energy	20 Kcal
Protein	0.7 g
Fat	0.2g
Carbohydrates	4.3 g
Fiber	10.1 g
Ash	0.4 g
Calcium	14.0 mg
Phosphorus	17.0 mg
Iron	0.8 mg
Thiamine	0.1 mg
Rivoflavin	0.03 mg
Niacin	3.8 mg
Áscorbic acid	15.0 mg
Vitamin A	7.0 mg

Source: Instituto de Nutrición de Centroamérica, 2012:50.

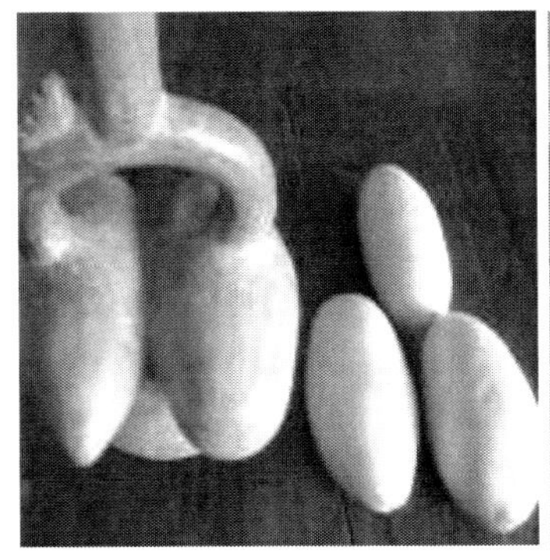

Figure 14. The tumbo serrano in Mochica ceramics. Source: imágenes de tumbo en internet and a plant in a garden. (Source: H. Córdova).

Tumbo Serrano (*Passiflora mollisima H.B.K*). Also known as "poroporo", "tumbito", "tumbito del monte", "tacso"; is a climbing vine of some five meters length, with fleshy roots. Its fruit is ellipsoid-like of 5 to 10 cm long and 3 to 5 cm diameter at its wider sector. It propagates by seeds and grows wild on fences and small trees. It is found in temperate environments with temperatures ranging from 14 to 22°C along the Andes from north of Chile to Colombia between 1000 and 3500 m altitude. In Peru it is cultivated mainly as fruit for household subsistence consumption. It grows well in the inter Andean valleys of Ancash, Junin, Moquegua and Huancavelica. This fruit has been cultivated since pre Hispanic times and it was known as "titín" or "apincoya" (Tapia y Fries, 2007: 115-118).

The tumbo fruit is a big Berry that holds many seeds wrapped in a mucilaginous sauer- sweet substance that resembles the passion fruit or maracuyá (*Passiflora edulis*). It is consumed fresh as fruit or in refreshments, jams, ice-cream, and others. It is also used as a basic ingredient to sauces; and the flowers are used to prepare tea to relax the nervous system. In traditional medicine it is appreciated by its therapeutic properties against kidney stones, urinary troubles, and stomach aches, among other uses.

The tumbo plant is native to the Central Andes (Wust, 2003:80) and even when it is being used since prehispanic times, it has not yet entered as a common market staple, especially in the larger markets of the coastal cities of Peru. It is an ideal fruit for summer refreshments because it is hydrant, low in calories but rich in minerals and vitamins, as it is shown in the following table.

Table 15. Nutritional composition of Tumbo Serrano (100 g of pulp)

Compound	*Amount*
Energy	64 Kcal
Water	82.1 g
Protein	1.2 g
Fat	0.5 g
Carbohydrates	5.4 g
Fiber	3.6 g
Ash	0.8 g
Calcium	8.0 mg
Phosphorus	34.0 mg
Iron	0.6 mg
Thiamin	0.02 mg
Riboflavin	0.11 mg
Niacin	4.56 mg
Ascorbic Acid	66.7 mg

Source: Tablas peruanas de composición de alimentos, 2009: 28-29.

Zarzamora (*Rubus robustus*, *R. fluribundus*, *Rubus sp.*). Also known as Cjari-Cjari (Cusco) o Siraca (Apurímac) is a small bush with many stems that depart from the soil level. Its stems may grow as high as three meters; they are flexible and covered by small curved thorns that extend even to the leaves and the inflorescence. Leaves are dark green upperside and whitish underside, trifoliated with sawing borders (Yacovleff and Herrera, 1933:137). Flowers are white and the fruit comes in racemes. They are ellipsoid berries 1.5 to 2.5 cm diameter and 3 to 5 g in wild state; are green when young and latter change to red and finally dark brown or black when they reach maturity. The flavor is sauer-sweet and may be eaten as fresh fruit or in jam, refreshment, ice cream, pastries, and others.

Once the plant starts to produce it continues generally with two harvests per year. Production starts after one year of planting and continues to 12 - 15 years before replacement.

The zarzamora grows best at temperatures between 12°C and 19°C, with 80% - 90% relative humidity, high sunshine and some 800 to 2500 mm annual rainfall well distributed to keep soil moisture. These ecologic conditions are well established in the Quechua region of Peru where this plant is widely distributed from 1500 to 3500 m altitude and according to latitude; at lower latitude it grows in lowest places and it keeps rising as latitude increases. Peruvians don't yet pay much attention to this fruit, but in the northern Andes, from Venezuela to Ecuador it is already being commercially cultivatedand; it is easily found at urban markets with attractive prices to the producers.

Figure 15. Zarzamora plant and fruit. (Source: H. Córdova).

Chapter 19

MEDICINAL PLANTS

Another topic of interest is the study of medicinal plants. The increasing interest on medicinal plants has to do with the search for remedies to cure the body and spiritual illnesses. This is a world attitude and to certain point, it is a challenge for any human society. In the Peruvian case there is long tradition that goes back to pre-Hispanic times, because the Andean societies not only experimented with food species but also with those that prevent and cure illnesses. Santiago Erik Antúnez de Mayolo (1990:40) notes that during the first years of Spanish colonization, when the Europeans could not cure an illness they went to the "Indians" looking for solutions. In fact, the Indians' food included nutraceutical species that provided them the nutrients for the body and the medicine by which they avoided any deficiency states or organic unbalances (*ibidem*, 29). Paying attention to the altitudinal environmental conditions where these species grow best, I include here the following table.

Achiote (*Bixa orellana L.)*. This is a shrub or small tree, of 1.0 to 6.0 m high. Its stem is woody, branchy and erect. The fruit is red and come covered by soft spines (Palacios 2006:20). This plant is native to the tropical climate between 100 and 600 m altitude, with annual precipitation of 1000 to 1500 mm and media temperature of 25°C to 30°C. It is known since pre Hispanic times and to present it is already a commercial crop. Besides its consumption as dye for food it has also medicinal applications in the treatment of respiratory affections (cough, bronchitis), as analgesic (headache), healing light wounds, against the leprosy, antidote (intoxication of cyanic acid or food containing it such as sauer manioc), burnings, laxative and digestive (Ibidem:22-23). Since recently it has called the attention as des-inflammatory of the urinary system and to curb prostatism. This is done by using the leaves, fruit and root.

Table 16. Medicinal plants according to altitudinal zones in the Central Andes 176

Common Name	Scientific Name	Used Part	Applications
Between 300 and 2000 m.a.s.l			
Achiote	*Bixa Orellana, L.*	Leaves, seeds, root	Cough, bronchitis, headache, healing light wounds, leprosy treatment, intoxication by cyan acid, hydric or food containing it like agria manioc; burning and laxative
Achira	*Canna indica L.*	Leaves, stem, root.	Local analgesic, dermal anti allergic, anti rheumatic and wound healing, anti coughing and mastitis, diuretic.
Ají	*Capsicum annun L*	Fruit	Digestive, cholagogue, carminative, overweight treatment, analgesic and rubefacient
Amor Seco	*Bidens pilosa L*	Leaves	Hepatic affections, halitosis, mouth affections, diuretic and scalds, emmenagogue and anti dysentery
Boliche	*Sapindus saponaria L.*	Root, bark & fruit	Anti hemorrhoids, anti varicose, haemostatic and anti-diarrheic; expectorant, diaphoretic and diuretic, varicose veins
Calahuala	*Polipodium angustifolium L*	Leaves and roots	Anti-inflammatory, anti rheumatic, anti diarrheic and haemostatic
Capulí Cimarron, Yerba del Chilalo	*Nicandra physaloides L.)*	Leaves and fruit	Diuretics, anti-coughing and anti-inflammatory
Chamico	*Datura stramonium L*	Dry leaves and fruit	Otologic and otalgia
Chanca Piedra	*Phyllantus niruri L*	Leaves, stem, root	To destroy calcium accumulations in the urinary system; diuretic, anti-inflammatory, galactogene; latex heals wounds, root is good against jaundice
Chilca	*Baccharis lanceolata Kth*	Leaves	Local analgesic, anti rheumatic, antispasmodic and fracture treatment

Common Name	Scientific Name	Used Part	Applications
Congona	*Peperomia galioides HBK*	Fresh leaves and stem	Antispasmodic; analgesic, fractures, cutting wounds, hepatic and heart illnesses
Drago, Sangre de Grado	*Croton palanostigma Klotzsch*	Resin	Antiulcer (gastric ulcer & gastrointestinal), anti hemorrhagic (light wounds), anti neoplasia (tumors), anti varicose veins, women urogenital illnesses, depurative, anti infection (throat affections), anti tuberculous and antiseptic (women genitals)
Paico	*Chenopodium ambrosioides*	Leaves, flowers	Anti helmíntic, digestive, anti diarrheic, anti malaria; anti inflammatory, anti rheumátic, anti parasit, and against flies bite.
Guanábana	*Anona muricata L.*	Leaves & root	Anti diabetes, anti spasmodic, Anti dysentery, antipyretic, anti neoplasia, Vulnerary
Guayabo	*Psidium guajava L.*	Fruit bark, leaves, root	Anti diarrheic, haemostatic, eupeptic, candidiasis and dental caries prevention
Hierba Santa	*Cestrum hediondinum Durn*	Leaves, flowers	Analgesic (sore throat, joints), digestive, emmenagogue; fresh leaves in water are good for baby bathing when they have rush-rush, anti pyretic
Huaranhuay	*Stenolobium mollis L.*	Leaves, flowers	Antiseptic, haemostatic (to cure light wounds), anti diarrhea and cutaneous affections (spots, insteps), skin bleaching; diuretic, sudorific.
Huito	*Genipa americana L.*	Fruit, seeds	Vaginal anti inflammatory, antitussive, treatment of jaundice, alopecia and urticaria; seeds are haemetic.
Lengua de Vaca	*Rumex crispus L.*	Leaves	Astringent (ocular irritation), laxative, analgesic, anti inflammatory; root is local antiseptic.
Matico	*Piper angustifolium RyP*	Leaves	Haemostatic, anti-inflammatory, dermatologic, genitourinary antiseptic, expectorant and anti jaundice.
Molle	*Schinus molle L.*	Young branches, leaves, fruit, resin	Respiratory affections (bronchitis & cough), hepatitis, depurative & antispasmodic; fruit against rheumatism, wound healing, and haemostatic.
Nogal	*Junglan neotropica D.*	Green fruit, leaves	Diabetes; non-infectious pharynx-tonsillitis (sore throat), dyed hair, hair loss, non-infectious diarrhea, non infected wounds
Pacae	*Inga feuillei D.C.*	Stem bark and root,	Anti hemorrhoid y anti bleeding (non infected light wounds); digestive,

Table 16. (Continued)

Common Name	Scientific Name	Used Part	Applications
		leaves, fruit pulp	gastrointestinal anti inflammatory; anti neoplasia (dermal and gastric cancer).
Quina, Cinchona	*Cinchona officinalis L.*	Bark and branches	Febrifuge, emmenagogue, astringent, antimalaria and vermifuge.
Sauce	*Salix humboldtiana Willd*	Leaves, bark	Anti inflammatory, anti rheumatic and analgesic; dysmenorrhea.
Tabaco	*Nicotiana tabacum L.*	Leaves	Antiwarty, anti rheumatic, anti asthmatic (pertussis) and otalgias.
Tara, Taya	*Ceasalpinia tinctoria HBK*	Fruit, leaves	Haemostatic, anti hemorrhoid and against non infectious diarrhea; non-infectious pharynx-tonsillitis (throat irritation)
Tumbo	*Passiflora quadrangularis L.*	Leaves, flowers, fruit	Sedative; analgesic, anti inflammatory, kidney stones.
Tuna	*Opuntia ficus Indica L.*	Leaf, fruit	Hypoglycemic, hypocholesterol, antitusigen, kidney depurative and anti inflammatory; antidiarrheic
Uña de Gato	*Uncaria tomentosa Willd*	Bark, root	Cancerostatic, anti inflammatory and antiulcerous; neoplasia preventive, anti rheumatism, depurative and diuretic.
Yacón	*Polymnia sonchifolia PyE*	Root	Fruit and food for diabetics
Between 2000 and 3500 m.a.s.l			
Culén	*Psoralea pubescens L.)*	Leaves and stem	Anti diarrheic, laxative, antidiabetic and antispasmodic
Marcco	*Ambrosia peruviana, Willd*	Root, stem, leaves and flowers	Antispasmodic, against colic anti rheumatic, anti inflamatory, neurotonic (invigorates the nervous system)). Flowers have worming applications.
Mocco-Mocco, Cordoncillo Rojo	*Piper stomachicum C.*	Warheads	Digestion and against rheumatic pain
Muña	*Minthostachys setosa Brig.*	Leaves and stem	Digestive (carminative), kidney affections, respiratory, analgesic, dermatomycosis, dandruf and antiseptic (wounds).
Mutuy	*Cassia hookeriana Gill*	Leaves	Laxative, diuretic and purgative.
Pinco Pinco	*Ephedra americana HyB*	Root, stem, leaves	Anti jaundice; against mouth infections, haemostatic, anti inflammatory and diuretic; diminishes blood pressure.

Common Name	Scientific Name	Used Part	Applications
Salvia	*Salvia oppositiflora RyP*	Leaves, flowers	Sore throat and inflammation; antitusigenous; antisweat and antispasmodic.
Sauco	*Sambucus peruviana HBK*	Leaves, flowers, root	Antitusigenous, diaforetic and anti rheumatic; anti inflammatory, diuretic; and the fruit is a mild laxative.
Valeriana	*Valeriana pinnatifida RyP*	Root	Antispasmodic, sedative of the central nervous system, against insomnia and contusions
Between 3500 and 4200 m.a.s.l			
Asmachilca	*Eupatorium gayanum Wedd*	Leaves, stem and flowers	expectorant and anti asthmatic
Escorzonera	*Perezia multiflora HyB*	Leaves and root	Antipyretic, sudorific, expectorant and diuretic
Hercampure	*Gentiana prostrata L.*	Whole plant	Colagogo, hipocholesterolemic, hepatic affections, obesity, normoglycemic, depurative and antiinfective
Huamanripa	*Laccopetalum giganteum Weed*	Flowers and leaves	Antitusigenous, expectorant
Maca	*Lepidium peruvianum Chacón*	Root	Antirachitic, antianemia, revitalizing, hormone alterations (goitre, sterility, menstrual cycle alterations). Natural viagra
Mitzca-Mitzca	*Geranium sessiliflorum Cav*	Leaves, stem	Antiseptic, haemostatic, to fight mouth afections
Pasuchaca	*Geranium dielsianum Knuth*	Whole dry plant	Hypoglycemic, antidiabetic, depurative, astringent atony, blennorrhagia, sores, diarrhea, pharyngitis, colds, bleeding hemoptysis, inflammations, menorrhagia, mouth and breast
Sanguinaria	*Oenothera rosea Ait*	Leaves, flowers, root	Vulnerary and ecchymosis cure (blows, wounds, contusions); antitusigenous; antirheumatism.
Yareta	*Azorella multífida RyP*	Stem and leaves	antirrheumatic, in myalgias and kinks

Source: Author's arrangement of data from Palacios, 2006; and Valdizán & Maldonado, 1922.

Achiote (*Bixa orellana L.).* This is a shrub or small tree, of 1.0 to 6.0 m high. Its stem is woody, branchy and erect. The fruit is red and come covered by soft spines (Palacios 2006:20). This plant is native to the tropical climate between 100 and 600 m altitude, with annual precipitation of 1000 to 1500 mm and media temperature of 25°C to 30°C. It is known since pre Hispanic times and to present it is already a commercial crop. Besides its consumption as dye for food it has also medicinal applications in the treatment of respiratory affections (cough, bronchitis), as analgesic (headache), healing light wounds, against the leprosy, antidote (intoxication of cyanic acid or food containing it such as sauer manioc), burnings, laxative and digestive (Ibidem:22-23). Since recently it has called the attention as des-inflammatory of the urinary system and to curb prostatism. This is done by using the leaves, fruit and root.

Figure 16. Achiote grains when ripe. (Source: H. Córdova and Internet).

Achira (*Canna indica L.*). A tall bushy herb of 1.5 to 3.0 m high. Its stem goes underground and its aerial sprouts are the parts we see above the surface. It has alternate leaves with a base extension that wraps the stem (Palacios 2006:25). This crop has extended along tropical America where it is cultivated up to 2000 m a.s.l (*ibidem*: 25) It is also used in traditional medicine as local analgesic, dermal anti-allergic, anti-rheumatic and healing light wounds. This is done by using the leaves, stem and root.

Amor Seco (*Bidens pilosa L.*). This is an annual herb of 20 to 60 cm high. Its leaves are deltoid, opposed and saw-like borders (Palacios 2006:44). It is found in agricultural fields on the Peruvian Coast and Yunga where it is seen as invasive plant. Its flowers and leaves are used to cure hepatic affections, halitosis, mouth affections, diuretic and also to eliminate scalds. It is also used as emmenagogo and anti dysenteric (*ibidem*: 45-46).

Figure 17. Achira. (Source: H. Córdova).

Asmachilca (*Eupatorium gayanum Wedd*). Shrub bush of 1.0 m high. Its leaves are linear, wrinkly and covered by a thick pilosity at their underside (*Ibidem*: 48). It is found at the western side of the Andes and at the inter Andean valleys of Peru, between the 3600 and 4000 m.a.s.l in Cusco, Puno, Apurimac and Ayacucho *(ibidem*). Its leaves, stems and flowers are used as expectorant and anti asthmatic (*ibidem*: 49).

Boliche (*Sapindus saponaria L.*). This is a branchy tree of about 10 m high. Its fruit is a dry berry, spheroidal that comes in racemes; when ripe turns to a yellowish or brown color. The seed is black bright (Palacios 2006:61). It is found in the western watersheds of the Peruvian Andes, in the warm and dry inter Andean Yungas and along the rivers of the coastal valleys (*ibidem*: 61). The fruit is used as expectorant, diaphoretic and diuretic; the root and bark is used to treat hemorrhoids, varicose veins and it is also haemostatic and anti-diarrheic (*ibidem*: 62).

Calahuala (*Polipodium angustifolium L.).* It is an herb of flexible stem. Its leaves are palmate like and extended hand, dark green or yellowish, depending on their exposition to sun (Ibidem: 71). It grows in places of temperate and humid climate. In Peru this species is found at the oriental side of the Andes as part of the humid forest of the eyebrow of the forest -Ceja de Selva- (*ibidem*: 71). Its leaves and roots are used as anti-inflammatory, anti rheumatic, anti-diarrheic and haemostatic (*ibidem*: 72).

Capulí Cimarrón, **Yerba del Chilalo** (*Nicandra physaloides L.*). This is an annual herb of about 1.0 m high. Its leaves are aovadas with saw-like borders. The fruit is rounded, like the *Physalis peruviana* and it encloses many small seeds in a dry environment. They are not edible. It grows well in hot and temperate environments, and it is widely distributed between the 200 and 2000 m altitude. The leaves and fruit are used in traditional medicine as diuretic, anti tussive and ant inflammatory (*ibidem*: 86)

Figure 18. A dry plant of yerba de chilalo. (Source: H. Córdova).

Chamico (*Datura stramonium L.)*. This is an annual herb of 0.8 to 1.2 m high. Its stem is semi-woody, and its leaves are oval, alternate and dented. When rubbing in hands expel an unpleasant odor. The fruit is a capsule ovoid, prickly and fleshy (*ibidem*: 95). The Datura genera are native to Peru and it is well distributed in tropical and subtropical environments at both sides of the Andes (*ibidem*). The dry leaves are used as antispasmodic; the leaves and fruit are utilized to control the otologic and otalgia affections (*ibidem*: 96)

Chancapiedra (*Phyllantus niruri L.*). This is an annual herb that generally grows in humid, shady places, with tropical climate. Its stem is erect of 30 to 60 cm high. Its flowers are small, isolated in the axis of each leaf. Its fruit are globular and oblate capsules of 2 to 3 mm diameter. The root is long with few branches *(ibidem*: 98). This plant is found in the upper forest –selva alta- region in the Eastern side of the Andes. Its leaves and stem are utilized to destroy the calcium cumulates in the urinary system; they are also used as diuretic, anti inflammatory and galactogen; the latex helps healing of wounds and the root is applied to treat the jaundice *(ibidem*: 100).

Figure 19. Chancapiedra. (Source: H. Córdova).

Chilca (*Baccharis lanceolata Kth.)*. This is a shrub of 1 to 1.5 m high. It has a very branchy stem that makes it like a bush. Its leaves are single, alternate and as spatula shape with sawlike borders *(ibidem*: 102). It grows in all the coast and sierra valleys of Peru, especially in humid soils and preferable acids from sea level to 2800 m.a.s.l. The leaves are used as local analgesic, anti rheumatic, antispasmodic and to arrange bone fractures (*ibidem*: 104)

Congona (*Peperomia galioides HBK)*. It is an erect herb of 20 to 45 cm high of pleasant odor. Its flowers are small and greeny as well as its fruit (*ibidem*: 126). It is widely distributed in tropical America and it is found in rocky places, non-exposed to winds and humid in the inter Andean valleys and also in the western sides of the Andes between 400 and 3000 m of altitude. The fresh leaves are utilized as antispasmodic; leaves and stem as analgesic, des-inflammatory, to arrange bone fractures, cutting wounds, hepatic and heart affections (*ibidem*: 127).

Figure 20. Chilca plant. (Source: H. Córdova).

Culén (*Psoralea pubescens L.*). This is an erect shrub up to 3 m high. Its leaves are trifoliate of dark green color. Its flowers are small and blue. Its fruit is oval, small and indehiscent. It grows in the Andean region between 2000 and 3800 m.a.s.l. The leaves are used to control diarrhea, and the whole plant is utilized as laxative, anti diabetic and antispasmodic (*Ibidem*: 135).

Drago, Sangre de Grado (*Croton palanostigma Klotzsch*). This is a tree that may reach up to 15 m high; almost always branchy at the upper part of the species. Its leaves are brad and aovads of 12 to 30 cm long. Its fruit is small, red and wrapped by hair (*ibidem*: 137). It grows well in the humid tropical forest from 125 to 2080 m.a.s.l; especially in the "purma" vegetation or area of forest regeneration after being cropped for five to eight continuous years in the oriental side of the Andes. Its geographic distribution is wide in South America and stretches along the Amazon basin and Central America. The resin is red and it is known as "sangre de drago" which is utilized against ulcers (gastric and gastrointestinal ulcers); it is also anti-bleeding (light wounds), anti neoplasic (tumors), anti varicose (varicose ulcers), uro-genital affections of women, depurative, anti infectious (throat affections), anti tuberculosis and antiseptic (women genitals) *(ibidem*: 139 -141)

Escorzonera *(Perezia multiflora HyB*). It is an herb of about 40 cm high, erect and flexible stem. Its leaves have the border double saw-like, dented and spiny (*ibidem*: 143). It grows in the arid and rocky land slopes of the lower Puna between 3800 to 3900 m.a.s.l, and strectches from Argentina to Colombia. Its leaves are used as antipyretic and sudorific; the leaves and root are utilized as expectorant and diuretic *(ibidem*: 144)

Guacatay *(Tagetes minuta).* Also known as "shilshil", "wacataya", "wacatea", is an annual herb of the Asteraceae family; erect that may reach some 50 cm high with flexible stem. It has lanceolate and dented green leaves with a strong odor. It is native to the Central Andes and it is widely known and cultivated to add special flavor to food sauces mixed with rocoto (hot pepper). However, it is also utilized in traditional medicine to help digestion, carminative and anti-abortion. The leaves as tea are recommended to alleviate gastric pain and the flowers and fresh leaves against colds and bronchitis. Its leaves also produce oil utilized in aromatherapy and perfumes.

Figure 21. Garden plant of guacatay. (Source: H. Córdova).

Hercampure (*Gentiana prostrata L.*). This is a bushy herb, small stem dark brown. Its leaves are small, simple and opposite; its rather purple flowers are also small. The fruit is a capsule with numerous seeds (Palacios 2006: 162). It grows between 3500 and 4300 m. a.s.l., at places very cold exposed to wind blowing. The whole species is used as colagogue, hypo cholesterolemic, hepatic affections, obesity, normoglycemic, depurative, and anti infection (*ibidem*: 163).

Hierba Mora (*Solanum nigrum*). This is an herb of one meter high that grows wild in agricultural fields and other open spaces with direct sunrays from sea level to 3000 m altitude. Its fruit are rounded berries of some 5.0 mm diameter, green that turn to black when ripe. They contain solanine which makes them toxic when eaten in big quantity. Its flowers are white. All the species is utilized to make emollient mixed with llanten (*Plantago officinale*), cola de caballo (*Equisetum sp*) and grama dulce (*Triticum repens*) to alleviate stomach inflammation after an alcoholic intoxication.

Hierba Santa (*Cestrum hediondinum Durn*). This is a shrub erect of 1 to 3 m high of unpleasant odor. Its stem is branchy from the base. Its leaves are alternate and oblanceolate petiolate. Its fruit is an small berry, oblong of 1 cm log, green that turns black or dark blue when ripe. It grows wild along the river sides and irrigation ditches in the coast and sierra between 200 and 3400 m a.s.l. The leaves and dry flowers are used as analgesic (throat sore, jonts), digestion and emmenagogue; the fresh leaves in water are utilized to bath babies when they have rush, antipyretic (Palacios, 2006: 166).

Huamanripa (*Laccopetalum giganteum Weed*). This is an erect herb some 50 cm tall. Its leaves are rosette with irregular lateral borders saw-like and dented in the upper half (Palacios 2006: 173). It is found on the karstic sideslopes of the Puna along the Andes between 4200 and 4600 m of altitude. Its flowers and leaves are utilized as antitussigen and expectorant.

Figure 22. Hierba Mora. (Source: Internet).

Figure 23. Hierba Santa. (Source: H. Córdova).

Figure 24. Huamanripa. (Source: Internet).

Huaranhuay (*Stenolobium mollis L.*). This is a small tree that may reach 6 m tall. Its stem is thick and cylindrical; Its leaves are compound, opposite and dark green with the upper side wrapped by hair (Palacios 2006:175). It is native to South America and grows wild in rocky soils and temperate climate. In the Andes, this species is found between 1600 and 3000 m altitude. Usually it is covered by flowers from November to March and from April to July at places near the coast. Its leaves and flowers are utilized as antiseptic, haemostatic (to cure light wounds), anti diarrheic and cutaneous affections (spots, instep), to decolorize skin; the flowers are used as diuretic and sudorific.

Huito (*Genipa americana L.).* This is a tree that may reach 20 m high. Generally it shows little foliage and the stem is conic shape and sometimes globose. Ovate leaves, opposite, with the upper side dark green and the underside light green. Its fruit is a berry ovoid of 6 to 12 cm long (Palacios 2006:178). It grows wild in the tropical and subtropical climate of the eastern side of the Andes from the Brazilian-Peruvian border to 3000 m altitude. The ripe fruit is utilized as vaginal anti inflammatory, anti tussigen, to treat jaundice, alopecia and urticaria. The seeds have an haemetic property.

Lengua de Vaca (*Rumex crispus L.*). It is a perennial herb of 40 to 90 cm tall. Its stem is erect and cylindrical with alternate green leaves. Its flower is very small, unisexual and bisexual. Its fruit is bright dark brown to brown (Palacios, 2006: 190). It grows wild in marshy soils from 400 to 4000 m of altitude. Its leaves are utilized as stringent (ocular irritation), laxative, analgesic and anti inflammatory; the root is good as local antiseptic

Maca (*Lepidium peruvianum Chacón*). This is a perennial herb of about 15 cm tall. Its root is a tuber, globose shape and rounded of 3 to 6 cm diameter. The stem is short and the rosette leaves try to get attached to soil to avoid the frosty temperature (Palacios 2006: 200). It grows in the high Andes between 3800 to 4500 m.a.s.l. It is cultivated on the Puna slopes and rather recently it has become attractive to the market due to its special qualities to increase man's virility. It is very resistant to cold and frost. Harvest is usually done in March and June. The root is utilized to cure rickets, it is anti anemia, restorative and to the treatment of hormone alterations (goitre, sterility, alterations of the menstrual cycle). For its energetic qualities it is considered as a natural Viagra.

Marcco (*Ambrosia peruviana Willd*). It is a perennial bush very fragrant. Its stem is branchy and tuberculous, with alternate leaves and aovad shape. Its fruit is wrapped by four prominences in tip (Palacios 2006:204). It grows wild in humid and marshy places, aside the rivers'borders on the coast, inter Andean and jungle valleys of Peru from 200 to 3500 m.a.sl. The whole species is used to curb rheumatism; it is anti inflammatory and neuro invigorator (invigorates the nervous system). The flowers have vermifuge applications.

Figure 25. Marcco shrub. (Source: H. Córdova).

Mastuerzo (*Tropaeloum majus L.*). This is a creeping and climbing herb, flexible, with length of near 3 m. Its leaves are smooth, green yellowish color; its flowers are big and come as bouquet in the leaf axis and the fruit is globose and indehiscent (Palacios 2006:226). It grows wild in any type of soil and climate in the Central Andes, from sea level to 3800 m of altitude. Besides, it resists environmental temperatures from 11 to 40°C (*ibidem*). Its flowers and leaves are used against the hair lose, dandruff, headache, wound healing, vulnerary, anti rickets, and anti scurvy. Moreover, the flowers are utilized to curb dermal affections (acne, spots, itching) and as antiseptic.

Matico (*Piper angustifolium R & P*). This is a perennial shrub, generally of some 2 to 2.5 m tall, but it may reach as 10 m high in primary forest. It has a knotty stem with numerous thin branches. Its leaves are simple, intense green and a characteristic pleasant odor. Its fruit comes in spike 6 to 10 cm long (Palacios, 2006:230). It grows wild from 400 to 2700 m.a.s.l, at both sides of the Andes; and it stretches from Bolivia to Central America. The leaves are utilized as homeostatic when there are wounds, anti inflammatory, dermatologic, antiseptic genitourinary, expectorant, and anti dysentery.

Mitzca-Mitzca (*Geranium sessiliflorum Cav*). This is a perennial herb. Its root is superficial and grows horizontally; the stem is relatively short and leafy; and the flowers are small and reddish (Palacios 2006:233). It is found wild at open spaces from 3800 and 4000 m a.s.l. Its leaves and stem are utilized as antiseptic, homeostatic and to cure mouth affections.

Mocco Mocco or **Cordoncillo Rojo** (*Piper stomachicum C.)*. It is a shrub of about 2 m high. Its stem shows pronounced knots; and its leaves are simple almost heart-like in the base. The leaf size is between 8 to 1o cm long (Palacios 2006:236). This shrub grows wild in shady places close to water streams in the sierra from 2500 to 2600 m.a.s.l. The leaves are used to ameliorate digestion troubles and against rheumatic pain.

Molle (*Schinus molle L.)*. It is a small tree of about 10 m high. Its stem is branchy and exudes a whitish resin when cutting the superficial bark layer. Its leaves are alternate and its flowers are small and numerous (Palacios 2006:239). It is a native species of Peru though there are other individuals distributed in the rest of America. It is found from 500 to 3300 m.a.s.l., and forms part of the typical vegetation of the Yungas. The young branches and leaves are used against respiratory affections (bronchitis and cough), hepatitis, depurative and antispasmodic; the fruit is good to cure rheumatism, light wound healing and haemostatic.

Muña (*Minthostachys setosa Brig.*). This is an herb of woody stem that reaches about 1 m high. Its leaves are somewhat rounded, light sawed or dented and very aromatic. The white flower comes at the leaves axis (Palacios 2006:243). It grows wild in the Andes from 2300 and 3400 m of altitude. The leaves and stem are used to control digestion (carminative), to treat kidney affections, respiratory troubles, analgesic, dermal mycosis, dandruff and antiseptic (wounds).

Figure 26. Molle tree. (Source: H. Córdova).

Figure 27. Muña. (Source: H. Córdova).

Mutuy (*Cassia hookeriana Gill*). It is a shrub of 3 to 5 m tall. Its leaves are compound, paripinnate and alternate; and aligned in spiral sequence, with upper side green-yellowish color and underside decolorized. Its flowers are yellow and grouped in bouquet as large as the leaves. Its fruit is a legume of 7 to 11 cm long and 1.5 to 2.0 cm wide, brown when ripe (Palacios 2006:248; Reynel & Marcelo, 2009). It grows wild in well drained places along the Andes from 2700 and 4000 m altitude. The leaves are utilized as laxative, diuretic and purgative.

Nogal (*Junglan neotropica D.*). This tree is native to America and is cultivated for its wood and medicinal properties. Its fruit is edible (Palacios 2006:251). It grows in the Andes from 100 to 3500 m of altitude and adapts quite well to humid places, especially at the Easter Andean side. The bark and leaves are utilized to prepare a substance to dye wool. In traditional medicine it is used the green fruit shell to cure diabetes; the leaves as tea are good to cure non infectious pharyngotonsillitis (throat pain), to hair dying and against hair losing, non infectious diarrhea, non infected wounds.

Ortiga *(Urtica dioica, Urtica sp).* It is a perennial herb of about 1.5 m tall. The stem is erect and quadrangular, and the leaves are about 15 cm long, deltoid with sawed borders, dark green with many urticant hairs. Flowers are bouquets on the leaves axis. It usually grows from sea level to 3800 m.a.sl., in organic soils well drained and open environment because it needs sunlight. This species is used dry or fresh to stimulate a good digestion; it is anti diarrheic, and helps hepatitis recovering. It is a mild laxative, diuretic, anti diabetes, and galactogen. Besides it is good to fight obesity, cholesterol, hypertension, gout, and joint problems. This species is very well known in traditional medicine due to its multiple properties.

Pacae (*Inga feuillei D.C.*). This tree was already described in section of edible fruit. Here, I will only mention the medicinal contribution to traditional knowledge. The leaves are utilized to make cataplasm to control diarrhea; the bark of root is anti hemorrhoids and antibleeding (non infected light wounds); the fruit pulp helps to get a good digestion. Besides, it is gastrointestinal anti inflammatory; the bark of stem is anti neoplasia (dermal and gastric cancer)

Figure 28. Ortiga. (Source: H. Córdova).

Paico (*Chenopodium ambrosioides L.*). It is an herb with strong aroma of 30 to 50 cm tall. Its leaves are oblanceolate and dented; and the flowers are scarce, greenish and small (Palacios 2006:261). It grows in warm and temperate environments on the rocky slopes and in cultivated fields from 100 to 3000 m of altitude. The leaves and flowers are utilized as anthelmintic, digestive, anti diarrheic, anti malaria, anti inflammatory, anti rheumatic, anti parasite and against insect bites.

Pasuchaca (*Geranium dielsianum Knuth*). It is a perennial herb, acaulescent. Its leaves are pubescent, membranous, palmate; its inflorescence is an umbel and the fruit is of regma type esquizocarpico (Palacios 2006:281). It is found from 3000 to 4300 m.a.s.l. The whole species is used as hypoglycemic, anti diabetic, depurative, atonias stringent, bleedings, sores, diarrhea, pharyngitis, colds, hemorrhagic hemoptysis, inflammations, menorrhagia, mouth and breast ulcers.

Figure 29. Dry Pasuchaca. (Source: Internet).

Pinco Pinco (*Ephedra americana HyB*). This is a perennial bush, erect or decumbent that may reach up to two meters high. Its flowers are opposite or whorled and the fruit is like a nut, reddish and acid flavor (Palacios 2006:287). It grows wild from 2400 to 4900 m.a.s.l. The root, stem, and leaves are against dysentery; it is also used in mouth affections, as haemostatic, anti inflammatory and diuretic; moreover it is effective to low blood pressure.

Figure 30. Pinco pinco. (Source: Internet).

Figure 31. Piñón. (Source: Internet).

Piñón (*Jatropha curcas L.*). This is a shrub that liberates translucent latex. Its leaves are lobed of 6 to 15 cm long and the fruit is a capsule of about 2.0 cm long; of light color with dark stretch marks (Palacios 2006:290). It grows mostly cultivated in warm and temperate climate from 80 to 500 m altitude. The grinded seeds are utilized as purgative, laxative and emmenagogue; the latex and leaves are used as anti inflammatory, and the leaves only are good expectorant and antitussive.

Quina or Cinchona (*Cinchona officinalis L.)*. Under this name there are a wide variety of trees and shrubs within the Rubiaceae family. The Cinchona is a tree of about 10 m high. Its dark gray trunk is of 1.0 to 1.5 m of diameter. Its leaves are opposite, lanceolated of 8 to 10 cm long and 3 to 4 cm wide. This species has a growth period very slow; and it is necessary to wait from 6 to 9 years to have the best harvest of medicinal interest. At this age the species has the greatest percentage of alkaloids in its bark (Palacios 2006:301). It is found in warm and humid climates; and its geographic distribution stretches from Bolivia to Venezuela. The bark and thin branches are used as febrifuge, emmenagogue, astringent, anti malaria and anthelmintic.

Figure 32. Cinchona or quina. (Source: Internet).

Salvia (*Salvia oppositiflora RyP*). It is an herb of woody stem that may reach some 60 cm tall. Its leaves are simple, deltoid and opposite; and its fruit is tetraachene. It grows in dry and rocky soils from 2300 to 4000 m.a.s.l, along the Central Andes. The flowers are used to sore throat and inflammation, antitussive; the leaves are anti sudorific and antispasmodic.

Figure 33. Salvia. (Source: Internet).

Sanguinaria (*Oenothera rosea Aiton*). It is a perennial herb of 30 to 100 cm tall. Its stem is ascendant and thin. Its leaves are irregularly dented of 2 to 5 cm long (Palacios 2006:313). It grows wild at ditches' borders and at the cultivated field borders from 3600 and 4000 m.a.s.l. Its geographic distribution is wide in America and it stretches from the Southwest USA to Bolivia (*ibidem*: 14). The leaves are used as vulnerary and to cure ecchymosis; the root is antitussive; and the flowers and leaves are utilized to cure rheumatism.

Sauce (*Salix humboldtiana Willd*). It is a tree of 10 to 15 m tall, the bark has a roughened surface and the branches are some-how decumbent. The fruit is a capsule with small seeds protected by thin hairs. It grows wild along the rivers sides from 400 to 2500 m of altitude; and stretches in South America from north Argentina and Chile to Venezuela. The leaves and bark are used as anti inflammatory, anti rheumatic and analgesic; the leaves are to control dysmenorrhoea.

Figure 34. Sanguinaria. (Source: Internet).

Sauco (*Sambucus peruviana HBK*). It is a tree that may reach 12 m tall. Its stem is convoluted of light gray color. Its leaves are bright green, compound; and it fruit is a berry that turns black when ripe with 5 to 6 seeds (Palacios 2006:320). It grows in the western watersheds and inter Andean valleys in temperate climate from 2000 to 3200 m a.s.l. The leaves and flowers are used to control cough; and they are also diaphoretic and anti rheumatic; the root and leaves are anti inflammatory, diuretic; and the fruit is a mild laxative.

Tabaco (*Nicotiana tabacum L.*). It is a woody herb that may reach 3 m tall, completely wrapped by viscose and sticky hairs. Its stem is cylindrical and branched at its upper sector. Its leaves are soft, alternate and its fruit is a capsule ovoid with numerous and small black seeds (Palacios 2006:324). It grows in tropical environments of America but its origin is in the Central Andes from 300 and 2500 m of altitude. In traditional medicine, they utilize the fresh leaves as antiwarts, anti rheumatic, anti asthmatic and in otalgia.

Figure 35. Tabaco. (Source: H. Córdova).

Tara or Taya (*Ceasalpinia tinctoria HBK)*. It is thick shrub or small tree with gray trunk wrapped by small spines; and it fruit is plain sheath, orange reddish with dark, small ovoid seeds. It is found in warm and dry places from 1300 to 2800 m of altitude. It stretches at both sides and in the inter mountain valleys of the Andes from Venezuela to Chile. The fruit is utilized as haemostatic, anti hemorrhoids and against non infectious diarrhea; the leaves and fruit are boiled and used in gargle against the non infectious pharynx- tonsillitis (throat irritation).

Figure 36. Tara tree with fruit. (Source: H. Córdova).

Tumbo (*Passiflora quadrangularis L.*). This species was already described above and here I only mention is medicinal properties. The flowers and leaves are utilized as sedative; and the root and fruit are used as analgesic, anti inflammatory and to attack the kidney stones.

Tuna (*Opuntía ficus Indica L.)*. It is a cactus of fleshy and prickly leaves that may reach 5 m tall. The leaves are palmate green stalks of 20 to 40 cm long and 20 cm wide. Its fruit is cylindrical and eatable. It grows wild in arid and rocky places from 300 to 2000 m altitude though actually it is part of the cultivated fruit and it may be bought at the urban markets. It resists dryness due to its capacity to keep water in the stalks and stems. The stalks or cladodios are utilized in traditional medicine as hipoglicemiante, hypocholesterolemic, antitussive, kidney depurative and anti inflammatory; its fruit are anti diarrheic.

Figure 37. Tuna stalk and fruit. (Source: H. Córdova).

Uña de Gato (*Uncaria tomentosa Willd)*. It is vine of 18 to 19 m long. When young it resembles and herbaceous and climbing plant. But later, it becomes woody and look- like it had a twist of its trunk. It has a pink stem and the leaves are reddish and opposite. Its fruit is gray of 3.5 to 4.0 cm long (Palacios 2006: 346-347). It is found in the Eastern side of the Andes, in the tropical forest from 300 to 800 m altitude. Its bark and root has great demand because it is cancerostatic, anti inflammatory and anti ulcer; the root is said to be very good to prevent neoplasic illnesses, and also is anti rheumatic, depurative and diuretic.

Valeriana (*Valeriana pinnatífida RyP)*. It is an herb of about 30 cm high with a characteristic odor. It has simple leaves of some 20 cm long and 6 cm wide. Its purple mottled fruit is plain of some 2.5 mm long (Palacios 2006:354). It grows in the inter Andean valleys from 2900 to 3000 m of altitude. Its root is used against spasm; it is sedative of the nervous systems, and it is good to combat insomnia and to alleviate contusions.

Yacón (*Polymnia sonchifolia PyE*). This is an herb with branchy roots from which grow cylindrical woody and hollow stems of 1.5 m tall. Roots are irregular, fleshy; and some of them may weight as 100 to 500g and measure up to 25 cm long. They are brown earthy externally, and the internal flesh is light orange of flavor somewhat like a cantaloupe. It grows cultivated from 900 to 3500 m altitude along the Andes. The root is consumed as fruit and food for diabetics

Yareta (*Azorella multífida RyP).* It is an elastic woody herb. Its stems develop at the root and give way to many plants. Its leaves are small, hard and adhere to the stem as flakes. Its flowers are very small and yellow. It grows in the high Andes of dry and cold climate from 3500 to 5000 m.a.s.l. From its stems is extracted the gum-resin and the leaves are used as anti rheumatic, to control myalgias and twists.

According to Herminio Valdizán and Maldonado (1922:247), in the first quarter of the XXth century there were a great variety of vegetal resources in the traditional pharmacopeia of the Peruvian population. For example, they noted that in Arequipa there was a very intense commerce of products like: linseed, copal, incense, alhucema, cebadilla, tara, matico, quinsaucho (mays type), melon seed, orejón de membrillo (quince), pasas, huarmimunachi, pupusa, toronjil, retama, peanut, frijoles, turpa, allpa-ticca, pepa de palta (avocado seed), pinco-pinco, airampo, yanalí, ratania, llantén macho, ortiga macho, inti-sunca, taco, chaco, llipta, chiuchi-piñi, recado, barro del mar, chaquiro, etc. In Cuzco, they sold among other products the pinco-pinco, sotoma, pupusa, allpa-ticca, pallares, frijoles of all types and colors, yanalí, cilantrus seeds, huito, sasahui, ccata, chuco.chuco, intisuncca, chachacoma, airampo, etc. In Lima they sold huamanripa, wirawira, ortiga macho, huanca.huasa, huamanpinta, and others.

Chapter 20

VALIDITY OF TRADITIONAL MEDICINE BASED ON PLANT RESOURCES

The deficient distribution of the health services in Peru obligates the population, especially those living at remote places, to trust in the healing power of plants. There, where there is not doctor or the pharmaceutical products are too expensive, there are plants, whose healing power have been transmitted to the families since remote times and continue alive from one generation to another. Some have acquired more knowledge than others and little by little they are becoming the healers locally known as "curanderos". The Andes offer the ecosystemic conditions to the existence of a great medicinal plants variety, to the point that some places have acquired especial names, such as the "cerro La Botica" located above the town of Cachicadán, capital of the district of the same name, in Santiago de Chuco province, Department La Libertad; other well known place is Las Huaringas in the sierra of the department of Piura, Province of Huancabamba, northwest Peru. Here the curanderism is mixed with esoteric elements associated to the forces of nature whose spirits habit isolated places, like mountains or inside the lagoons.

Then, the proposal is that we must take advantage of these environmental conditions that well incorporated in the cultural patterns of householders may help to improve their means to get an extra income going into business. This means to start doing agriculture of the wild species that have some value as food or medicine. In fact, there are some plants that are already considered to have nutraceutic properties, that is, they give nutrients as well as chemical elements that protect the body against illnesses. It is necessary to cultivate those wild species and stop being collectors only. Agriculture will allow an expansion of the resources and secure their surviving in front of the climate change that is already previewed in the near future. By doing this, we also buy some time to continue doing experimental analysis to know the chemical compounds of those species used to cure many illnesses not yet solved.

A bird's view to the Central Andes shows the validity of the curanderos or chamans and the number of stores selling processed products based on plants well known by the local houselholders. In communal markets it is very popular to find stands selling medicinal herbs without any previous treatment used to cure many body and spiritual illnesses. A good compilation of these plants and the ways how to use them was made by the physicians Hermilio Valdizán and Ángel Maldonado (1922), followed by others like David Werner, Carol Thuman and Jane Maxwell with their publication *Donde no hay doctor: una guía para los campesinos que viven lejos de los centros médicos* (1973, 1995, 2010); and others. Some samples given by Valdizán and Maldonado continue being valid in spite of the time passed, and are commented below.

Bronchitis is treated eating onion (Arequipa) or radish (Lima, Huánuco, Junín, Ancash (1922:247). Pains in the lower sector of stomach, specifically in the intestine are alleviated with an emollient elaborated with fragments of quince, orange and pineapple (1922: 256).

Joint rheumatism is treated by frictions with the roasted fruit of molle (Puno) and topical applications of wheat grains heated previously (Cuzco) (1922: 259). In Arequipa, they use pink geranium cooked to cure the smallpox (1922: 264).

The wart or Carrion illness is controlled in Lima by frictions to the affected part of the body with light roasted altamisa and complete baths with a cooking of willow, arrayán, and cat's claw. To disappear the eruptions of wart, it is advisable to make local applications with grinded maguey seeds or tomato cimarron seeds. To favor the wart eruption and its quick destruction they use the following: infusion of the quishuar leaves or trunk, infusion of molle, infusion of elder, infusion of willow, different types of wine, water of mote (cooked corn), infusion of bear's herb, guarapo, infusion of suelda con suelda, light roasted sugar cane, infusion of cat's claw, infusion of huachauso and infusion of mincausho (1922: 269).

The stomach colics are treated in Apurimac with quinoa and coca macerated in sugar cane brandy (1922: 270). As carminative are utilized the infusions of camataipaco and jeto-jeto (Junin), of acayusa, cidrera and ajenjo (Loreto); and of sallika (Cuzco). As digestive are employed the infusions of beet, pirgumuna, and panizara (Cajamarca) and abuta-bitter (Loreto) (1922: 272).

The external treatment of leishmaniasis in Huarochiri is made through the use of lemon juice, ground quina or cinchona, walnut, yarabisca, dog foot, llanten, and others (1922: 291). The mucosal leishmaniasis -espundia- is treated in Junin and in Puno with wheat; in Ayacucho with the latex of a squash leaf; and in Arequipa with the herb called swallow *(1922:293-5).*

The banana bark is used in Arequipa to cure calluses; with the same end are used in Huanuco the tomato; and in Ica, the smashed garlic (1922: 296-7). Scabies are attacked in Cajamarca with the cream made of apple (1922: 298).

In some departments of southern Peru, people do the cooking of chick peas roasted and honeysuckle leaves to facilitate the childbirth. Moreover in some departments of north Peru, people do for the same thing, a maceration of amaryllis in sugar cane brandy; and in Cuzco they make the parturient to catch with her hands the hot cobs of fresh corn (1922: 340).

The cooked rue is well known to be abortive in all Peru. The same attributes are assumed to the cooked lecrae. At some places in South Peru, people cure uterine deviations by rubbing on the woman womb hot cloth with the rest of the cooked quince fruit (1922: 342). To recuperate the normal menstrual cycle they make a substance of cooked male nettle, and the infusion of hualhua, hinojo, mancapaqui, llantén, culantrillo, and equisetum (Arequipa), milma-sacha (Lambayeque), celery, feverfew, parsley, cooked garlic leaves, enolado of tutapaya leaves and cooked onion

(Cuzco); infusion of grape leaves, chamomile flowers and culantrillo of well (Lima); cooked date seeds (Ica) infusion of manayupa (Junin); infusion of roasted chick peas and matico, cooked barley, cumin, muña, mutuy and madre-quisa (departments of the South) (1922: 343). The lacteal secretion is increased in Cuzco by massages with quinoa or drinking corn chicha or an infusion of anis (1922: 344-5). The babies' umbilicus healing is accelerated by pouring matico powder, cascarilla, burn totora and burn cotton (1922: 348).

Smallpox is controlled with the retama flower and corn powder (Junin) (1922:362). The whooping cough is treated with sugar cane honey, garlic honey and red tuna (Arequipa); and to carry lemon necklaces (Lambayeque (1922: 363).

Neuralgias are cured in Huacho and other people of the Coast with a rubbing of a substance based on verbena and matagusano; in Apurimac they employ the itana herb, burned hot pepper and the tumbo flower; in Cuzco, they utilize the quisccamati and the chapiquiscca, as well as the sauco flower. On the same way, in Cajamarca, Lima, Huanuco, Junin, and Puno they put slices of oca and olluco on the patient's forehead to treat the headache; it is also applied topically the coca leaves at the head temples of the patient (1922: 384-5).

The epilepsy and the hysteria are controlled in Cuzco by chewing coca leaves. Insomnia is controlled by applying on the forefront some lettuce leaves impregnated with olive oil (Arequipa). In Junin they put eucalyptus leaves. Finally, in the same department, they make the patient to drink hot chichi de jora. In Lambayeque, they put floripondio under the pillow (1922: 388-90).

Alcoholism is combated in Junin with huachangana infusions (1922: 392). To heal small cutting wounds they apply the ancco-cebadilla juice and leaves of light baked yerba santa (Cuzco); the smashed leaves of piri-piri or the chamairo (found in the forest), jattaco leaves (found in the mountains), molle resin (departments of South Peru). In almost all the country they use the burned cotton. In Loreto, they use the llausaquiro latex; and in northwest Peru it is used the powder of a fungus called oreja de palo (1922: 404-6). To consolidate the bones, people use leaves of rosemary and quinoa, and chuño, wheat, tarwi, bull's gall and catholic balsam (Arequipa); also wheat and chamana leaves (southern departments); leaves of maqui-maqui, miyua, and pingacu-sacha (from the jungle) (1922: 409).

Dislocation is attended by using the tincture of hualtaco and enolado of huambuquero (northwest Peru); and in Junin, they make the patient to drink the recanancha (1922:410).

Hemorrhoids are cured with washes of a maceration made of lettuce root and cabbage (Arequipa) (1922: 417). The stomach bleeding is cured with the cooking of liga, male llanten, equisetum and bolsa-bolsa (departments of South); the same is employed with the cooked siempre-viva (Piura); and in Loreto it is used the resin of sangre de grado. In Cajamarca, they combat the uterine bleedings with tapatapa (1922: 422-23). Other species cited by Valdizan and Maldonado are listed in the following table:

Table 17. Common and scientific names of vegetal species and types of medicinal use

Common Name	***Scientific Name***	***Utilization***
Achicoria	*Cichorium intybus L*	Depurative.
Achupalla	*Puya sp*	Against ear pain and anti scurvy
Aitacopa	*Hedyosmum racemosum G. Don*	Against rheumatism
Ajo	*Allium sativum L*	Against fever, scaby, instep, chilblains, calluses and insect bites
Alcaparrillo	*Cassia sp*	Purgative
Alfalfa	*Medicago sativa L*	Against cough, and hair tonic
Alfilercillo	*Erodium sp*	Blenorrhage.
Aliso	*Alnus jorullensis HBK*	To decrease the milk secrecion
Ajonjolí	*Sesamum indicum L*	Galactofore, against fractures, dislocations and contusions

Table 17. (Continued)

Common Name	*Scientific Name*	*Utilization*
Almendro	*Prunus amygdalus Stokes*	Against hemorrhoids
Altamisa	*Ambrosia peruviana Willd*	Against rheumatic pain, leg cramps and to hemorrhoid des-inflammation
Ambarina	*Scabiosa atropurpurea L*	Ocular des-inflammation and skin illnesses
Amor seco.	*Bidens sp*	Against urinary retention, hepatitis and hidropesia
Anis	*Pimpinella anisum L*	Carminative and galactofore
Apio	*Apium graveolens L*	Against pneumonia, rheumatism, and as digestive stimulant
Árbol del pan	*Artocarpus incisa Forst*	Against the mouth illnesses
Arroz	*Oryza sativa*	Stomach desirritant, dysentery, anti diarrheic, and to combat acne
Arveja	*Pisum sativum L*	Against measles, smallpox,
Aya-aya	*Alonsox acutifolia RyPav*	To kill teeth pain
Ayapana	*Eupatorium triplinerve Vahl*	Sudorific
Azucena	*Lilium candidum L*	To facilitate child birth
Balsamina	*Momordica balsamina L*	Against contusions, ulcers and as vulnerary
Barbas de Maíz	*Zea mais*	To reduce feet swelling especially of pregnant women
Berro	*Nasturtium fontanum Aschers*	Against the lung's inflammation, tonsillitis, constipation, and depurative
Bichayo	*Capparis avicenniifolia HBK*	Epilepsia treatment, anti goitre, anti spasmodic, and hypnotic properties

Common Name	***Scientific Name***	***Utilization***
Bolsa-bolsa	*Capsella bursa pastoris*	To cure contusions
Borraja	*Borrago officinalis L*	Sudorific and against measles
Buenas tardes	*Mirabilis japala L*	Diuretic and purgative
Cafeto	*Coffea arabica L*	Against pneumonia, diabetes and the hiccup
Caigua	*Cyclantera pedata Sehrad*	Against otitis and tonsilitis
Cala	*Zantedeschia aethiopica Spreng*	Teeth anti inflammatory
Camote	*Ipomoca batatas Lam*	Anti inflammatory of contusions and galactofore
Campanilla	*Bomarca sps*	Anti haemorrhagic
Campanillamorada	*Pharbitis sp*	Against insomnia
Cáncer-ccora	*Senecio sp*	Ulcers treatment
Canela	*Cinnamomum zeylanicum Breyn*	Anti diarrheic
Cañafistula.	*Cassia fistula L*	Laxative
Cañihua	*Chenopodium sp*	Against the spider bite, against disentry
Cardo santo	*Argemone mexicana L*	Against whooping cough, anti athsmatic, febrifugue, laxative, and invigorator
Carrizo	*Arundo donax L*	Diuretic and analgesic against the scorpium bite
Ccarpunya	*Piper Carpunya RyPav*	Colds and pneumonia
Cebada	*Hordeum sativum Jessen*	Used to combat measles, smallpox, and anti inflammatory
Cebadilla.	*Schoenocaulon officinale*	Againts lices
Cebolla	*Allium Cepa L*	Against insomnia, epistasis, chilblains, angina pectoris, to facilitate chilbirth, urinay illnesses and against pnumonia
Cedro	*Cedrela fissilis Vell*	Abortive
Cetico	*Cecropia sp*	Against the strongest

Table 17. (Continued)

Common Name	*Scientific Name*	*Utilization*
Chachacuma	*Escallonia resinosa Pers*	Brain invigorator and carminative
Chamana	*Dodonacea viscosa L*	Dislocations, bone fractures
Chechecra	*Senesio sp*	Against malaria
Chicche	*Lepidium sp*	Against bleeding
Chicmu.	*Trifolium sp*	Vulnerary
China-huanarpo	*Jatropha basiacantha Pax*	Afrodisiac
Chinchilcuma	*Mutisia viciaefolia Cav*	Cleaning wounds.
Chirimoyo	*Anona cherimolia Mill*	Cefalalgia and disentery
Choloque	*Sapindus saponaria L*	Stringent, invigorator, and to fix fractures
Chuchu-chuchu	*Baccharis genistelloides Pers*	Against malaria and rheumatic pain
Chupa sangre	*Xylopleurum roseum Raim*	To absorb the blood
Cidro	*Citrus medica L*	Carmitative, relaxing and against fungus
Ciruelo agrio	*Spondias purpurea L*	To erase smallpox scars
Cinamono	*(Melia azedarach L)*	Rheumatism and fever
Cipres	*Cupressus sempervirens*	The fruit is restrainer when applied to the bleeding
Clavelina	*Malesherbia sp*	Against asthma
Coca	*Erythroxylon coca Lam*	Headache, stomach pain, colic, anti diarrheic, chilblains, and anti rheumatic
Col	*Brassica oleracea L*	Against bronchitis, and decrease the milk production on women
Comino	*Cuminum Cyminum L*	Carminative
Conoc	*Senecio mathewsii Wedd*	Against rheumatism
Conuca	*Werneria sp*	Antiheemorrhagic
Copal	*Hymenaca sp*	Against headache

Common Name	*Scientific Name*	*Utilization*
Cucharilla	*Embothrium grandiflorum Lam*	Anti ulcer, against the affections to uter and against teeth pain
Culantro	*Coriandrum sativum L*	Carminative.
Desflemadera	*Spilanthes sp*	To cure scuvy
Escobilla del Perú	*Scoparia dulcis L*	Febrifugue and y astringent
Escorzonera	*Perezia multiflora HyB*	Diuretic, sudorific, febrifugue, expectorant and against measles
Esparrago	*Asparagus officinalis L*	Against urinary retention and kidney stones
Espino	*Opuntia sp*	Hepatic treatment
Eucalipto	*Eucaliptus globulus Labill*	Against asthma
Flor de muerto	*Tagetes patulus L*	Against asthma and whooping cough
Frejol	*Phaseolus vulgaris L*	To fix fractures and to normalize menstrual cycle
Fresa	*Fragaria vesca L*	Anti scuvy, eliminate kidney stones and hepatic affections, as depurative, astringent, disentery and blenorrhagia
Garbancillo	*Astragalus garbancillo Cav*	Fracture alleviation
Garbanzo	*Cicer arictinum L*	Blenorrhagia, against smallpox and malaria
Gigantón	*Trichocereus cuzcoensis Britton y Rose*	Against hydrophobia, wounds, and diuretic
Genciana	*Gentiana sp*	Febrifuge and emmenagogue
Grama dulce rama-kachu	*Cynodon dactylon Pers*	Depurative and diuretic
Granadilla	*Passiflora liguralis Juss*	To prevent yellow fever, cough, paludism, digestive and febrifuge

Table 17. (Continued)

Common Name	*Scientific Name*	*Utilization*
Granado	*Punica granatum L*	Anti diarrheic, anti disentery, mouth ulcers and anti abortion
Guaco	*Mikania guaco HBK*	Against snake bite, poisonous animals and anti rheumatic
Guindo	*Prunus avium L*	Against disentery
Habilla.	*Hura crepitans L*	Purgative and anti asthmatic
Hancahuasa	*Senecio rhizomatus Rusby*	Against acne, pneumonia and vulnerary
Higuera	*Ficus Carica L*	To cauterize calluses, fractures, to reduce hernias, to disappear spots, anti casp, anti diarrheic, and against hydropesia
Higuerilla	*Ricinus communis L*	Desinflamation of spleen, haemorrhoids, purgative, fever, and against blenorrhagia
Hinojo	*Foeniculum vulgare Mill*	Carminative y galactofore
Hualtaco	*Loxopterygium huasango Spruce*	To treat dislocations
Huamampinta	*Chuquiragua huamanpinta Hieron*	Diuretic and anti blenorrhagic
Huaranccaiso	*Ranunculus sp*	Against nasal bleeding, vulnerary and rubefacient
Huaynacuri	*Valeriana sp*	Antispasmodic and anti rheumatic
Huicho	*Cyclantera sp*	Purgative.
Humuto	*Alternanthera sp*	Relaxation and purgative.
Incienso	*Boswellia carteri Birdw*	Broken bones
Inti-suncja	*Tillandsia usneoides L*	Anti irritant (almorrans), heart illness, lung and liver illness

Common Name	*Scientific Name*	*Utilization*
Jattacco	*Ustilago maidis Lev*	The spores are utilized as haemostatic powder and dried for small cutted wounds
Jasmin	*Jasminum grandiflorum L*	Carminative, some neurosis and rheumatic pains
Jengibre	*Zingiber officinale Roscoe*	Anti hemorrhagic
Juaisch	*Vallea stipularis L*	Astringent in the eyes illnesses
Juan Alonso Espina de perro	*Xanthium spinosum L*	Against hepatic affections, of the stomach, diuretic, against kidney illnesses, spleen and ovary
Lechuga	*Lactuca sativa L*	Sudorific
Lengua de vaca	*Rumex patientia L*	Treatment of many dermal illnesses, to cure wounds, astringent in the eyes and depurative
Liga	*Psittacanthus sp*	To consolidate fractures.
Limero	*Citrus limetta Risso*	Against headache and carminative
Limonero	*Citrus limonum Risso*	Against the whooping cough, dandruf, bite of poisonous animals, calluses, diarreas, rheumatic affections, and angina
Lino	*Linum usitatissimum L*	Against constipation, hydropesia, and burns
Llamac-ñahui	*Mucuna rostrata Benth*	Vermifugue tenifugue
Llantén	*Plantago major L*	Astringent, wound lavative, analgesic, wounds, blood bleeding, and against the hemoptisis of the lung
Madreselva	*Lonicera confusa DC*	Sudorific and diuretic
Magnolia	*Magnolia grandifora L*	Against colic
Maguey	*Agave americana L*	To consolidate fractures, to close wounds, to reduce swellings, anti stomach

Table 17. (Continued)

Common Name	*Scientific Name*	*Utilization*
		inflammatory, against hepatic illnesses, against apendicitis hydrophobia and against the eyes inflammation
Malvita de olor	*Pelargonium odoratissimum Aiton*	Against fleas
Mamey	*Mammea americana L*	Against lices
Manzanilla	*Matricaria chamomilla L*	Carminative and to favor childbirth
Manzano	*Malus domestica*	Anti dandruf and hemorrhoids
Marañón	*Anacardium occidentale L*	Astringent
Margarita	*Polianthes tuberosa L*	Nervous affections
Mastuerzo	*Tropacolum majus L*	Against face insteps and eye.inflammation; as anti inflammatory, anti scurvy and headache
Mata gusano	*Flaverin contrayerva Person*	Clean wounds and against cough
Melocotonero	*Prunus pérsica L*	Purgative and against albinurria
Membrillejo	*Cordia rotundifolia RyPav*	Against jaundice
Membrillo	*Cydonia vulgaris Pers*	Against disentería, anti diarrheic and to cure the eyes
Milenrama	*Achillea millefolium L*	Astringent.
Mito	*Carica candicans Gray*	Digestive
Mogo-mogo	*Piper moho-moho C*	Vein pain
Mosqueta	*Bahuinia sp*	Purgative
Mucle	*Encelia canescens Cav*	Galactofore
Muña.	*Satureja boliviana Briq*	Carminative
Nabo	*Brassica napus L*	Against cough and bronchitis

Common Name	*Scientific Name*	*Utilization*
Naranjo	*Citrus aurantium L*	Headache, spider bite, sedative of nervous, diarrea and antispasmodic
Ñorbo	*Passiflora sp*	Against disentería,
Occa-occa	*Oxalis sp*	Trush child
Oje	*ficus sp*	Against anchilostoma, anti helmintic
Olivo	*Olea europea L*	Against spiders and scorpion bite, against constipation, scabies, liver stones, and urinary retention
Olluco	*Ullucus tuberosus loz*	Erysipela and migraine
Ortiga	*Urtica urens L*	Against cramps, migraine, colic, and nasal bleeding
Ortiga macho	*Cajophora*	Epistasis, pneumonia and to cure sciatica
Paico	*Chenopodium ambrosioides L*	Vermífuge, astringent, against skin crust and anti diarrheic
Pájaro bobo	*Tessaria integrifolia RyPav*	Against asthma
Pallar	*Phaseolus pallar Molina*	Against eye inflammation and smallpox
Palmera de dátil	*Phoenix dactylifera L*	Re establish menstrual cycle
Palmera de coco	*Cocos nucífera L*	To stop uterine bleeding, tenifugue and galactofore
Palto.	*Persea gratissima Gartn*	Diabetes, astringent, disentery, diarrhea and antiofidic
Papayo	*Carica papaya L*	Digestive and to treat anginas
Papelillo	*Bougainvillea spectabilis Willd*	Against pneumonia
Papilla	*Euphorbia sp*	Purgative
Pega-pega.	*Boerhaavia sp*	Jaundice
Pensamiento	*Viola sps*	Depurative and anti scrofulous
Pepa de cedrón	*Simaba cedron Planch*	Against snake bite

Table 17. (Continued)

Common Name	*Scientific Name*	*Utilization*
Perejil	*Petroselinum sativum Hoffm*	Stop nasal bleeding, to reduce milk secretion, against paludism and mosquitos bite
Pero	*Pirus communis L*	Anti constipation
Pez colofonia	*Pinus sp*	Lung pain
Pimienta negra	*Piper nigrum L*	To tap caries and stimulate circulation
Pimpinela	*Fragaria chiloensis Ehrh*	Astringent
Piñón	*Jatropha curcas L*	Purgative and for constipation
Plátano	*Musa paradisiaca L*	Against shaking chills, against calluses, to close small wounds and constipation
Porotillo	*Capparis sp*	Diuretic.
Pura-pura	*Wernerаria sp*	Against chloridric dispepsia and digestive
Quinua	*Chenopodium quinoa Willd*	Throat anti inflammatory, to facilitate childbirth, to recover bone fractures, galactofone, sudorific, against tyfoidea, laxative, purgative, and fever
Quishuar	*Buddlela globosa Lam*	Astringent and against warts
Retama	*Spartium junceum L*	Against rheumatic pain, headache, jaundice, and diuretic
Romero	*Rosmarinus officinalis L*	Invigorator of nervous system and brain, carminative, anti spasmodic, and against pneumonia
Rosa de jerico	*Anastatica hierochuntina L*	To facilitate childbirth
Rosa de remedio	*Rosa centifolia L*	Migraine, to facilitae chilbirth, laxative, intestinal

Common Name	*Scientific Name*	*Utilization*
		desinflammatory, and anti hemorrhoids
Ruda	*Ruta graveolens L*	Dismenorrhea, abortive, vermifuge, antispasmodic and against epylepsia
Sábila	*Aloe vera*	Against poison in general, purgative, apendicitis, migraine, ulcers, anti irritant, against angina and asthma
Salvia	*Salvia sp*	Antirrheumatic and carminative.
Sanango	*Tabernaemontana Sananho RyPav*	Against rheumatic pain and syphilis
Sano-sano	*Alsophila sp*	The stem exudes a glutinous materia utilized as vulnerary to cure wounds and chancre. It is also used as astringent
San Pedro	*Trichoreus pachanoi*	To treat infected wounds and skin illnesses. To clear water
Sauce	*Salix humbodltiana Willd*	To extirp eyeclouds, astringent in uterine cleaning, anti rheumatic, anti diarrheic, digestión stimulant, ulcers and anti dandruf
Siempreviva	*Helichrysum orientale Tourn*	Analgesic
Solimán	*Lobelia decurrens Cav*	Caustic and purgative
Sotoma	*Perezia coerulescens Wedd*	Diuretic and diaforetic
Tola	*Lepidophyllum quadrangulare Benth y Hook*	Against dysenteria
Tomate	*Solarum Lycopersicum L*	Calluses, against tonsillitis, laringitis, and against piles
Toñuz	*Pluchea sp*	Snake bite
Toronjil	*Melissa officinalis L*	Stomachal, antispasmodic

Table 17. (Continued)

Common Name	*Scientific Name*	*Utilization*
Toronjo	*Citrus decumana L*	Nervous affections
Totora enea	*Typha domingensis Pers.*	Wound healings of head and belly buttom
Tumbo	*Passiflora quadrangularis L*	Eliminate kidney stones and against cough
Tuna	*Opuntia ficus indica Mill*	Cough treatment, alleviate erisipelas and to cure hidrofobia
Tutumo	*Crescentia cujete L*	Wounds suppuration and against asthma
Ulux	*Columellia obovata RyPav*	stomachal and against intermitent fever
Varita de San José	*Wernería sp*	Antiasthmatic and pectoral.
Verdolaga	*Portulaca oleracea L*	Dysentery, emmenagogue and vermifuge
Verbena	*Verbena bonariensis L*	Clean ulcer rebel
Violeta	*Viola odorata L*	Ulcers cancerous
Yahuar-Chchuncca	*Oenothera multicaulis RyPav*	Against swelling
Yarabisca	*Jacaranda acutifolia HyB*	Astringent and vulnerary.
Yareta	*Azorella sps*	Antirrheumatic
Yerba bolsilla	*Calcecolaria sp*	Diuretic and against hydropesia
Yerba buena	*Mentha sp*	Dentrific, desirritant, carminative, antispasmodic and antidiarrheic
Yerba del Alacrán	*Heliotropium synzystachyum RyPav*	To treat wounds and to clean the head
Yerba del gallinazo	*Chenopodium opulifolium Schrad*	Insomnia, jaundice and indigestion
Yerba de la purgación	*Boerhaavia scandens L*	Depurative, antisyphilic and purgative

Common Name	*Scientific Name*	*Utilization*
Yerba luisa	*Andropogon schoenanthus L*	Carmitative
Yerba mora	*Solanum nigrum L*	To expulse worms from the caried teeth
Zanahoría	*Daucus carota L*	Diuretic and emmenagogue.
Zapallo	*Cucurbita máxima Duchens*	To eliminate scars, against pneumonia, sudorific, tenifugo and against otitis

CONCLUDING REMARKS

RURAL PERIPHERIES: PROBLEMS AND OPPORTUNITIES FOR DEVELOPMENT IN PERU AND THE PLANT'S DIVERSITY

In 1994, M. Redclift wisely noted that the sustainable development concept was too closely defined by the experiences, scientific theories and economic practices of the industrialised countries for it to be as attractive as it ought to be for states and peoples in the developing world. The "green" aspirations and hopes present in the former – which work to prompt society into saving resources, protecting the environment, limiting the consumption of non-renewables, gaining a more pro-environmental education and so on, are all based on an understanding that the natural environment is in and of itself a value to be prized by humankind (Henseling 1995). There are many developing countries in which the intellectual elites take the same view. However, the political and economic decisions taken at different levels of the administration unfortunately make it clear how very much under pressure the natural environment is, given the conviction that development and an improvement in living conditions can only be achieved through actions seeking to make the fullest possible use of resources. Likewise, bearing in mind the non-beneficial social and political practices that accompany such activity (not least corruption), the consequence is invariably the devastation of the environment, not merely from the point of view of nature and wildlife, but also when it comes to human beings. Just a few examples of this type of phenomena include large pools of crude oil in northern Ecuador, the contamination by mercury of waters and soils in northern Peru, and the heavy-metal pollution present in the rivers of the Peruvian Andes.

Local communities are under no illusions as regards the threats that pollution of the natural environment can pose, and so strive to fight against new developments (sometimes even successfully). Unfortunately, their efforts do not prove effective in most cases (as with the women of Mazahua, central Mexico, who fought to preserve intact and uncontaminated a body of water that supplied indigenous people with both drinking water and irrigation water for their fields (Skoczek 2013). *In extremis*, local people are even forced to abandon lands that have been occupied by their people and families for centuries.

In turn, in the urban and industrial spatial systems that have taken shape in countries of the global North since the Industrial Revolution, inhabitants are aware of threats arising from the pollution of the environment and the limiting of their life space. For this reason, movements in the name of ecodevelopment and environmental protection are a strong motor force behind development policy there. A comparison of the approaches to resources and valuable features of the natural environment in the developed and developing countries shows that the thinking and economic activity in the former have brought about a shift in the priorities and values that mainly speak for "development", as long as this takes place in a "sustainable" manner.

When it comes to the criteria of peripherality presented above, all are met by the region in the north-western part of the Peruvian Andes selected by us to serve as an example.

1. To begin with the simplest (geographical) criterion, it can be noted how the region is at the edge of the country, not far from the border with Ecuador. It is thus literally peripheral in respect of the centre, which is the Peruvian capital Lima. Peripherality thus

denotes a great distance from the centre, and hence a marginal location.

2. The consequence of such a location – far from the capital and in a natural environment quite different from that characterising even the seat of the regional authorities (with Piura located on the Pacific Lowland) – is neglect as regards education and environmental protection. Doctors are absent, and the level of education is very low, *inter alia* because there is also a shortage of teachers.
3. A poor system of infrastructure (most especially as regards roads) makes it difficult to incorporate local producers into global supply chains, and even to have them meet requirements on the domestic market as regards on-time deliveries, freshness and continuity of supply.

The further conclusion arising from the analysis is that circumstances of shortages and shortfalls combined with a low quality of life to hinder the incorporation of sustainable development principles into the local economy. While the slogan is currently popular among local politicians and authorities, it is seen to mean very little in practice. Only once a certain level of development is achieved, as well as increased awareness and an entrenchment of the idea at middle levels (including also in agriculture) will it be possible to commence with a discussion on the introduction of sustainable development principles. At any earlier stage, this will just be a slogan that inhabitants will fail to respond to.

The Andean biodiversity offers to the world a great variety of vegetal species to secure the human livelihood. Only in the Central Andes have been identified near 25,000 species of which 17,143 are flowering plants. There have been domesticated 128 native species, some of which with thousands of varieties, such as the potato, that actually is one of the most important crops of the world aside rice, corn, and wheat. At present there are 6,034 native species under human utilization in Peru; and from those there are 710 food species and 1,109 medicinal plants (Brack Egg and MendiolaVargas, 2000).

Here I only showed you some of the wild species of interest, some of which are in the process of incorporation to agriculture because of their inherent conditions that make them to have a good potential to help to solve the nutritional deficiencies, the health and ecological attributes for the Peruvian society. The climate variations force us to take measures to confront it and one of this is reforestation. The question is always on their lips: what species to choose? And here we have passed through some of them thinking in the benefit maximization, that is, plants that may offer wood, fruit and shadow to catch water vapor and keep temperature as low as possible. These native species are also germoplasm banks to which recur when looking for segments of native species to get more resistant crops to any illness and to make them more productive at industrial scales (Antúnez de Mayolo, 1981; Tapia, 1993). Then, it s necessary to give more attention to wild species, in terms of knowledge and evaluate their food and nutraceutical abilities to propose a better management, avoiding their extinction by over extraction in wild state.

REFERENCES

Aguirre, E. & Castillo, P. (2009). Extracción y estudio comparativo de las enzimas proteolíticas del fruto toronche (*Caricastipulata* y de la papaya (*Carica papaya*) y su aplicación en la industria alimentaria. Guayaquil, Ecuador, Facultad de Ing. Mecánica y Ciencias de la producción. *Escuela Superior Politécnica del Litoral*. Online, 31/12/13 ["Extraction and Comparative Study of Proteolytic Enzymes of Toronche Fruit (*Caricastipulata* and Papaya (*C. papaya*) and their Utilization as Food" industry]: http:// www.dspace.espol.edu .ec/ bitstream/123456789/ 7532/1/Extraccion%20y%20Estudio%20Comparativo%20de%20las%20Enzimas%20Proteol%C3%ADticas.pdf.

Ayala, G. (1992). Aporte de los cultivos andinos a la nutrición humana. In: FAO: *Raíces andinas-Contribuciones al conocimiento y a la capacitación*. Roma. ["Contribution of Andean Crops to Human Nutrition" in FAO: *Andean Roots- Contribution to their Knowledge and Training*, Rome].

Barrera, V., Tapia, C. & Monteros, A. (Editores). (2004). Raíces y tubérculos andinos: Alternativas para la conservación y el uso sostenible en el Ecuador. Serie: Conservación y uso de la biodiversidad de Raíces y Tubérculos Andinos: Una década de investigación para el desarrollo (1993-2003). N° 4. Instituto Nacional Autónomo de Investigaciones Agropecuarias. Centro Internacional de la Papa,

Agencia Suiza para el Desarrollo y la Cooperación. Quito, Ecuador-Lima, Perú. ["Roots and Andean Tubers: Alternatives to their conservation and sustainable uses in Ecuador". Series: *Conservation and Uses of Andean Roots and Tubers Biodiversity: One Decade of Research for Development (1993-2003)* N° 4. National Autonomous Institute of Agricultural Research. International Potato Center].

Blog "El tumbo: fruto de la pasión". (2012). ["Tumbo: Fruit of Passion"] Unknown author in: http://wwweltumbo.blogspot.com/, 11/12/13.

Brack, Egg A. & Mendiola, V. C. (2000). *Ecología del Perú*. Lima; PNUD, Editorial Bruño. [*Ecology of Peru*].

Britton, N. L. & Rose, J. N. (1919). *"The Cactaceae: Descriptions and Illustrations of Plants of the Cactus Family."* Washington, D.C; The Carnegie Institution of Washington. *Vol. 2.*

Cárdenas, M. (1969). Manual de las plants económicas de Bolivia. Cochabamba, Bolivia Imprenta Icthus. [*Handbook of Economic Plants of Bolivia*].

Córdova-Aguilar, H. (2013). *Nuevo Plan de desarrollo local concertado del distrito de Frías, provincia de Ayabaca*, Piura, 2013-2025. Lima, Municipalidad Distrital de Frías- Sociedad Geográfica de Lima. [*New Concerted Local Development Plan for the District of Frías, Province of Ayabaca*, Piura, 2013-2025. Lima, District Municipality of Frías-Geographic Society of Lima].

Córdova-Aguilar, H. (2014). Nuevo Plan de desarrollo local concertado del distrito de Viques, provincia de Huancayo [New Concerted Local Development Plan for the District of Viques, province of Huancayo], Junín, 2014-2025. (unpublished).

Czerny, M. (red.). (2013). Bieda i bogactwo we współczesnym świecie. *WUW*, Warszawa. [Poverty and wealth in the modern world].

Czerny, M. (1986). Planes del desarrollo regional y la integración del espacio en América Latina. *Actas Latinoamericanas de Varsovia, vol. 2,* 89–102. [Regional development plans and spatial integration in Latin America". *Latin American Acts of Warsaw. vol. 2*: 89–102].

Czerny, M. (1985). Działalność obcego kapitału a regionalne dysproporcje w Ameryce Łacińskiej. In: Neokolonializm. *PAN* Warszawa. [Activity of foreign capital and regional disparities in Latin America].

Czerny, M. (1980). *Urbanizacja w warunkach rozwoju zależnego na przykładzie Kolumbii.* [*Urbanization in conditions of sustainable development, the case of Colombia*]. Materials from the symposium: "Rozwój zależny krajów Trzeciego Świata". Kozubnik.

Czerny, M. & Córdova Aguilar, H. (2014). Livelihood – Hope and Conditions of a new Paradigm for Development Studies. *The Case of Andean Regions*. NOVA, New York.

Czerny M., Czerny A. 2002. The Challenge of Spatial Reorganization in a Peripheral Polish Region. *European Urban and Regional Studies* vol. 9, no. 1, s. 60–72.

De Feo, V., Simone, de F., Arias Arroyo, G. & Senatore, F. (1999). Carica candicans Gray (Mito), an Alimentary Resource from Peruvian Flora. *Journal of Agriculture and Food Chemistry, Vol. 47, N° 9*, 3682 -3684.

Elliott, J. A. (2013). (Fourth Edition). *An Introduction to Sustainable Development. Routledge Perspectives on Development.* London and New York.

Escobal, J. & Valdivia, M. (2004). *Perú: Hacia una estrategia de desarrollo para la sierra rural.* Lima, Grade. [*Peru: Towards a Development Strategy for the Rural Sierra*, Lima Grade].

Ewards, R. M. (1992). Development – Aufgabe des politischen Ordnungsrahmens oder ethisch motivierte unternehmerische Initiative? In: Sustainable Development als Leitbild der umweltbewußten Unternehmensführung. *Dokumentation einer Vortragsveranstaltung am, 22*, Juni 1992, Westfälischen Wilhelms-Universität zu Münster.

Felles, D. (2013). Diversidad biológica, cultural, y Vasconcellea candican, "mito", como especie subutilizada, en el distrito de Pachangara-Oyón-Lima. Universidad Nacional Faustino Sánchez Carrión, Repositorio Digital. [Biologic, Cultural Diversity and *Vasconcellea candican,* "mito" as underused species in the Pachangara-Oyón-Lima District". National University Faustino Sánchez Carrión. Digital Repository].

Fondo Internacional de Desarrollo Agrícola. (2013). Dar a la población rural pobre del Perú la oportunidad de salir de la pobreza. FIDA. ["*Give the Rural People the Chance to Leave Poverty*". FIDA].

Francis, J. K. (1994). "Inga fagifolia" (L.)Willd.Guamá. *SO-ITF-SM-72.* New Orleans,

LA: U.S. Department of Agriculture, Forest Service, Southern Forest Experiment Station.

Franco, G., et al. (2002). "Manual técnico - El cultivo del Lulo." ["*Technical Manual- The Lulo Crop*"]. Manizales, Colombia, Asohofrucol, Corpoica y Fondo Nacional de Fomento Hortofrutícola, Agosto.

Fritz, P., Huber, J. & Levi, H. W. (1995). *Nachhaltigkeit in naturwissenschaftlicher und sozialwissenszaftlicher Perspektive.* S.Hirzel, Wissenschaftlich Verlaggesellschaft, Stuttgart. [*Sustainability in a scientific and socio-scientific perspective.* S.Hirzel, Scientific publishing house, Stuttgart.].

Handke, K. (1993). Pojęcie "region" a symbolika "środka". In. K. Handke (red.): Region, regionalizm – pojęcia i rzeczywistość. *Instytut Slawistyki PAN*, [The term "region" and the symbol "center". In. K. Handke (ed.): Region, regionalism - concepts and reality. *Institute of Slavic Studies].* Warszawa, pp. 105-120.

Hauff, V. (ed.). (1987). Unsere gemeinsame Zukunft. *Der Brundland-Bericht der Weltkommission für Umwelt und Entwicklung*, Greven. [Our common future. *The Brundland Report of the World Commission on Environment and Development*, Greven].

Held, M. & Geißler, A. (Hrsg). (1993). *Ökologie der Zeit. Vom Finden der rechten Zeitmaße* [*Ecology of time. From finding the right time measurements*], S. Hirzel Verlag, Stuttgart, Leipzig.

Henseling, K. O. (1995). Eine nachhaltig zukunfts-verträgliche Stoffwirtschaft als politisches Leitbild. In: Fritz P., Huber J., Levi H.W. (red.). *Nachhaltigkeit in naturwissenschaftlicher und sozialwissenschaftlicher Perspektive.* S. Hirzel – Wissenschaftlicher Verlagsgessellschaft, Stuttgart. [A sustainably future-compatible material economy as a political model. In: Fritz P., Huber J., Levi

H.W. (Red.). *Sustainability in a scientific and sociological perspective*. S. Hirzel - Scientific publishing company, Stuttgart].

Hernández Bermejo, J. E. & León, J. (ed.). (1992). Cultivos marginados: Otra perspectiva de 1492. Roma, Colección FAO: *Producción y protección vegetal*, N° 26. ["Marginal Crops: Another Perspective of 1492", Rome, FAO Collection: *Vegetal Production and Protection* N° 26].

Hermann, M. (1992). Andean Roots and Tubers: Research Priorities for a Neglected Food Resource. [*Andean Roots and Tubers: Research Priorities for a Neglected Food Resource*] Lima, IPC.

Huamán Leandro, L. R. (2013). "Proyecto de tesis sauco final." ["Thesis Project Sauco Final."] Huánuco, Universidad Nacional Hermilio Valdizán. http:// es.slideshare.net/ LicethRocioHuamanLea/proyecto-tesis-sauco-final Page visited on December 17, 2016.

Instituto de Nutrición de Centro América y Panamá (2012). Tabla de composición de alimentos de Centro América. Guatemala, Unión Panamericana de Salud; Tercera reimpresión. [*Table of Nutritional Content of Food in Central America*].

Jiménez-Escobar, N. D., Estupiñán-Gonzalez, A. C., Sánchez-Gómez, N. & Garzón, C. (n.d). Entobotánica de la media montaña de la serranía del Perijá." ["*Ethnobotany of Medium Landslopes of the Perija Mountain*"]. http:// www.academia.edu/ 15298959/ETNOBOT% C3% 81NICA_DE_LA_MEDIA_MONTA%C3%91A_DE_LA_SERRAN% C3%8DA_DEL_PERIJ%C3%81 Consulted on December 17, 2016.

León, J. (1964). Plantas alimenticias andinas. Instituto Interamericano de Ciencias Agrícolas. *Boletín técnico N° 6*, Lima. ["Andean Nutrition Plants". Panamerican Institute of Agricultural Sciences. *Technical Bulletin N° 6*, Lima].

Lim, B. & Spanger-Siegfried, E. (red.). Co-authored by Ian Burton, Elizabeth Malone, Saleemul Huq. (2004). *Adaptation Policy Frameworks for Climate Change: Developing Strategies, Policies and Measures*. UNDP.

Lucas, Urquillas, K. A., Maggi Tenorio, J. M. & Yagual Chang, M. J. (2010-2011). Creación de una empresa de producción, comercialización y exportación de tomate de árbol en el área de Sangolquí, Provincia de Pichincha. Guayaquil, Tesis de Grado para el título de Ingeniería Comercial y Empresarial. Escuela Superior Politécnica del Litoral, Facultad de Economía y Negocios. ["Starup of Business on Production, Marketing and Export of Tree Tomato in the Sangolqui, Pichincha Province" Guayaquil. Thesis to get the Professional Title of Marketing Engineering].

Ministerio de Agricultura y Desarrollo Rural. Cadena Productiva Frutales y su Industria (2008). *Proyecto: Identificación, valoración y uso potencial de las pasifloras en el Huila con fines de mercados especializados del orden nacional e internacional*. Colombia. Corporación Centro de Investigación para la Gestión Tecnológica de Passiflora del Departamento del Huila - CEPASS Huila. Universidad de Huila. ["*Project: Identification, Valuation and Potential Use of Passiflorae in the Huila with the Object to Sell at Specialized Markets at National and International Order*"]. http:// www.huila.gov. co/ documentos/ agricultura/ CADENAS%20PRODUCTIVAS/ IDENTIFICACI%C3%93N,%20VALORACI%C3%93N%20Y%20USO%20POTENCIAL%20DE%20LAS%20PASIFLORAS%20EN%20EL%20HUILA%20CON%20FINES%20DE%20MERCADOS%20ESPECIALIZADOS%20DEL%20ORDEN%20NACIO(1).pdf.

Ministerio del Ambiente. (2010). *Dirección General de Calidad Ambiental Estudio Plan de Calidad ambiental Perú-Ecuador: Catamayo-Chira y Puyango-Tumbes*. Lima, Consorcio E & H. [*Study of the Plan of Environmetal Quality Peru-Ecuadro: Catamayo-Chira and Puyango-Tumbes*. Lima].

Ministerio de Salud. (2009). Instituto Nacional de Salud. *Tablas peruanas de composición de alimentos*. Lima. Octava edición, 70 p. [*Peruvian Tables on Nutrient Content of Food*].

Mayolo A. & de, Santiago E. (1988). *La nutrición en el antiguo Perú*. Lima; Banco Central de Reserva del Perú. Fondo editorial. [*Nutrition in Ancient Peru*].

Mayolo, A. & de, Santiago E. (1990). Uso del espacio agrícola precolombino. In: El mundo andino en la época del descubrimiento.

Lima, Comisión Nacional Peruana del V Centenario del Descubrimiento Encuentro de Dos Mundos. *CONCYTEC.*, pp. 21 -108. [Uses of Precolombian Agricultural Space". In *Andean World at the Time of Discover*, Lima].

Muñoz Fonseca, J. R. (1997). *La pitahaya.* [*The Pitaya*] Versión online: www.liberia. co.cr/promo/pitahaya_files Visitada el 4/6/2004.

Muñoz Jáuregui, A. M., Blanco Blasco, T., Alvarado Ortiz-Ureta, C., Serván, K., Ramos Escudero, F., Laja, B. & Navarrete Siancas, J. (2005). *Estudio nutritivo, bioquímico y toxicológico del fruto de la Carica stipulata V.M. Badillo (Papayita olorosa). Horizonte Médico, Revista de la Facultad de Medicina de la USMP*, Vol. 5, N° 2, pp. 57-61. *["Nutritional, Biochemical, and Toxicologic Study of Carica stipulata V.M Badillo"*].

National Research Council. (1989). *Lost crops of the Incas. Little known plants of the Andes with promise for worldwide cultivation.* Washington, D.C. National Academy Press.

Nolasco, C. D. & Guevara, P. A. (2009). Estudio de las principales características físicoquímicas y comportamiento del sanqui (*Corryocactus brevistylus sub sp. Puquiensis (Rauh & Backberg)* Ostolaza en almacenamiento. *Ancient, UNALM, 70* (4). [Study of Principal Physicochemical Characteristics and Behavior of Sanqui (*Corryocactus brevistylus sub sp. Puquiensis (Rauh & Backberg) Ostolaza* in Warehouse" *Ancient,* UNALM 70 (4).]

ONU, (2013). *Objetivos de desarrollo del milenio*: Informe de 2013. Nueva York. [*Development Objectives of theMillenium: Inform 2013*].

Palacios, J. (2006). *Plantas Medicinales Nativas del Perú.* 3rd ed. Lima: Imprenta ALC. [*Native Medicinal Plants of Peru*].

Pulgar Vidal, J. (1987). *Geografía del Perú: las ocho regiones naturales.* Lima. Ed. PEISA. [*Geography of Peru: The Eight Natural Regions.* Lima].

Petzold, H. (1997). *Nachhaltigkeit und "neuzeitlicher Städtebau" – zur kulturellen Dimension der nachhaltiger Stadtentwicklung.* IÖR-Schriten. [*Sustainability and "modern urban development" - the cultural dimension of sustainable urban development.* IÖR-Schriten.].

Prebisch, R. (1959). Commercial policy in the underdeveloped countries. *American Sociological Review*, v. *49*, no. 2, pp. 251–273.

Redclift, M. (1994). "Reflections on the 'Sustainable Development' Debate". *International Journal of Sustainable Development & World Ecology*, *1*, pp. 3-21.

Reynel, C. & Marcelo, J. (2009). *Árboles de los ecosistemas forestales andinos. Manual de identificación de especies*. Lima. Serie Investigación y Sistematización N° 9. Programa Regional Ecobona – Intercooperation. [*Trees of the Andean Forest Ecosystems. Manual of Species' Identification, Lima*].

Rodríguez, R. & Peña Segura, J. O. (1984). *Flora de los Andes*. Departamento Nacional de Planificación. Corporación Autónoma Regional de las Cuencas de los ríos Bogotá, Ubaté y Suárez. Colombia. [*Flora of the Andes*. National Department of Planning.]

Romero Rodríguez, E. (2012). *Desarrollo sostenible. Hacia la sostenibilidad ambiental*. Ed. Produmedios, Bogotá. [*Sustainable Development. Towards Environmental Sustainability*.]

Roorda, N. (with P. Blaze Corcoran, and J.P. Weakland). (2012). *Fundamentals of Sustainable Development*. Routledge. London and New York.

Roszkowska-Mądra, B. (2009). *Koncepcje rozwoju europejskiego rolnictwa i obszarów wiejskich*. [*The concept of development of European agriculture and rural spaces*]. gospodarkanarodowa.sgh. waw.pl/ .../ gospodarka_narodowa_2009_10_0.

Rościszewski, M. (1974). Przestrzeń krajów Trzeciego Świata – problemy metodologiczne. In: M. Rościszewski (red.). Przestrzeń krajów Trzeciego Świata. *Przegląd Zagranicznej Literatury Geograficznej, zeszyt*, 1-2. IGPAN, Warszawa. [The space of the Third World – methodological problems. In: M. Rościszewski. *The space of the countries of Third World*].

Sagástegui Alva, A. (1995). *Diversidad florística de Contumazá*. Trujillo – Perú, Editorial Libertad., 203, p. [*Floristic Diversity of Contumazá*. Trujillo].

Sagástegui Alva, A., Rodríguez Rodriguez, E. & Arroyo Alfaro, S. J. (2007). "Plantas promisorias: El mito o papaya silvestre." [Promisory Plants: Mito or Wild Papaya"]. *INNOVA NORTE*, *1*(1), pp. 109-119.

Sánchez, I. (1992). Frutales andinos. In: *Cultivos marginados, otra perspectiva de 1492*. Colección FAO, Producción y Protección vegetal N°26. Roma, Italia. ["Andean Fruits". In *Marginal Crops, Another Perspective of 1492*. FAO Collection. Vegetal Production and Protection N° 26. Rome].

Santos, M. (1971). La spécificité de l'espace en pays sous-développés: quelques aspects significatifs. *Institut d'Etude du Développment Economique et Social*, *Doc. De Travail*, no. 28. [*Space Identification in Under Developed Countries: Some Significant Aspects*].

Schneidewind, U. (1993). Ökologie und Wettbewerbsfähigkeit: Freiräume für eine nachhaltige Chemiepolitik nutzen. Diskussionspapiere. Institut für Wirtschaft und Ökologie. Hochschule St. Gallen Nr. 11. [Ecology and Competitiveness: Using Freedom for a Sustainable Chemical Policy. Discussion Papers. Institute of Economics and Ecology. University of St. Gallen No. 11].

Soukup, J. (1970). Vocabulario de los nombres vulgares de la flora peruana. Lima, Edit. Salesiana S.A. [Vocabulary of Folk Names of the Peruvian Flora. Lima].

Skoczek, M. (red.). (2013). *Zmiany społeczno-gospodarcze w regionach tubylczych Meksyku i ich percepcja przez mieszkańców. Przypadek Regionu Mazahua*. [*Socio-economic changes in the indigenous regions of Mexico and their perception by the inhabitants. The case of Mazahua region*]. WGSR UW, Warszawa.

Yi-Fu-Tuan. (1987). *Przestrzeń i miejsce* [*Society and Space*]. PIW, Warszawa.

Tapia, M. E. (1993). Semillas andinas. El Banco de oro. Lima; *CONCYTEC*., 76, p. [Andean Seeds. The Golden Bank. Lima].

Tapia, M. E. (2000). *Cultivos andinos subexplotados y su aporte a la alimentación.* Santiago De Chile, FAO, Segunda Edición. [*Underused Andean Crops and their Contribution to Food.* Santiago de Chile]

Tapia, M. E. & Fries, A. M. (2007). *Guía de campo de los cultivos andinos*. Roma-Lima, FAO –ANPE. [*Field Guide of Andean Crops*. Rome].

Trivelli, C., Escobal, J. & Revesz, B. (2009). *Desarrollo rural en la sierra. Aportes para el debate*. Lima, IEP, CIPCA, GRADE, y CIES. [*Rural Development in the Sierra. Contributions to Discussion*. Lima].

Van Den Eynden, V., Cueva, E. & Cabrera, O. (1998). *Plantas silvestres comestibles del sur del Ecuador*. Quito, Abya Yala. [*Wild Edible Plants in Southern Ecuador*. Quito].

Villegas Nava, P. (2012). *Los recursos naturales de Bolivia*, Cochabamba, Centro de Documentación e Información Bolivia-CEDIB, Segunda edición. [*Natural Resources of Bolivia*. Cochabamba].

Valdizán, H. & Maldonado, A. (1922). La medicina popular peruana. Lima, *Torres Aguirre*., *3* vol. [Peruvian Popular Medicine. Lima].

Vite, Vigo. & Andrea, A. (2014). Agricultura de plantas silvestres en la sierra de Piura: Análisis de un modelo productivo de cultivo y comercialización de pitaya rn Frías (Ayabaca, Piura). Tesis de Licenciatura en Geografía y Medio Ambiente, Pontificia Univrsidad Católica del Perú. [Agriculture of Wild Plants in the Sierra of Piura: Analysis of a Productive Model to Crop and Commercialize Pitaya in Frias (Ayabaca, Piura). Lima. Thesis of Professional Licence in Geography and Environment, Pontifical Catholic University of Peru].

Werner, D., Thuman, C. & Maxwell, J. (2010). Donde no hay doctor: una guía para los campesinos que viven lejos de los centros médicos. Berkeley, California (USA), Hesperian. [Where there is not a Physician: A Guide to Peasants Who Live Faraway of Medical Centers. Berkeley].

Wust, W. H. (2003*). Guía de especies útiles de la flora y fauna silvestre*. Lima, Diario La República; ediciones PEISA S.A.C, Primera edición. [*Guide of Useful Species of the Wild Flora and Fauna*. Lima].

Yacovleff, E. & Herrera, F. (1933). *El Mundo Vegetal de los Antiguos Peruanos, Cusco.*, *160*, p. Online: www.fortunatoherrera.tk. [*The Vegetal World of Ancient Peruvians*. Cusco].

Zapata Cruz, M. (2007). Caracterización bromatológica del frutal silvestre Mote-mote Allophylus mollis (Kunth) Raldkofer (Sapindaceae). ["Bromatollogic Characterization of the Wild Fruit Mote-Mote *Allophylusmollis* (Kunth) Raldkofer (Sapindaceae)"]. *Arnaldoa, Trujillo*, *14* (1), pp. 105-110.

Zerka, P. (2015). Obywatele zasobni w zasoby. Przesłanki, wzorce i instytucje dla zrównoważonego rozwoju w Polsce. Tezy i propozycje do "Białej Księgi" zarzadzania zasobami naturalnymi. demos Europa, Fridtjof Nansen Institute. Warszawa. [Citizens resource-rich. Premises, patterns and institutions for sustainable development in Poland. Theses and proposals to the "White Book" of management of natural resources].

ABOUT THE AUTHORS

Miroslawa Czerny
Full Professor
University of Warsaw,
Department of Geography and Regional Studies
Email: mczerny@uw.edu.pl

Prof. Dr. Dr. H.C. Miroslawa Czerny - MA 1974, University of Warsaw, Doctor 1978, University of Warsaw, post-doc (hab.), 1995 University of Warsaw. The title of professor of Earth Science 22.X. 2007, currently ordinary professor UW. 1976 (10 months) scholarship of Polish Government - University of Havana, Cuba 1978/1979 (1 year) technical assistance contract Instituo Geografico "Agustin Codazzi", Colombia 1986 (4 months) visiting professor, Universidad Autónoma del estado de Mexico, Toluca 1990-1992 Alexander von Humboldt post-doc scholarship - University of Tübingen, Germany 1992 (2 months) Europa scholarship of the Alexander von Humboldt Foundation in Oxford University (st. Antony's College). 1994 (1 semester visiting professor) University of Mainz. 1997 (6 months) PHARE scholarship, University of Oxford, School of Geography. 1997 (1 semester visiting professor) University of Mainz, College in Germersheim. Repeatedly visiting professor (lectures, courses, laboratories) at the University of Castilla-La Mancha, Complutense de

Madrid, at the Autonomous University of Barcelona (Spain). Coimbra (Portugal), Pontificia Universidad Católica del Peru (Lima), Universidad Autonoma Metropolitana Unidad Xochimilco, Universidad Nacional del Estado de Mexico, Toluca, Universidad en Jalapa (Mexico), Universidad de Manizales, Universidad del Rosario, Universidad Javeriana, Universidad de los Andes, Universidad del Valle, Cauca (Colombia), Universidad Nacional del Sur, Bahia Blanca, Universidad de Generel Sarmiento, Buenos Aires (Argentina), the Universidade de Caxias do Sul (Brasil), University of Indiana, Bloomington (USA) - among others. Numerous scientific Project, among other during last years: "Livelihood (the security of existence) - hope and conditions of a new paradigm for development studies", "Strategies for promoting sustainable rural development in regions with high levels of poverty. The concept of research methodology applied to mountain region in Northwestern Peru", "Old and new in the cities in Latin America - the actors and arguers of changes". More than 200 publications – articles, and books. Dr. H.C. on the Universidad Nacional de San Agustin in Arequipa (Peru).

Hildegardo Córdova Aguilar
Director, Centro de Investigación en Geografía Aplicada (CIGA-INTE)
Pontificia Universidad Católica del Perú
Email: hcordov@pucp.edu.pe

Dr. Hildegardo Córdova Ahuilar: - University of Wisconsin- Madison, 1976-1982; Ph.D in Geography, 1982. - Universidad Nacional Mayor de San Marcos 1964-1968; B.A. in Geography (1969) and Doctor in Geography, (1980). - University of Texas at Austin, 1970-1972; M.A. in Geography, 1972 - Bachelor of Arts in Geography, 1964-1968. University of San Marcos, 1969.

Major Field of Interest:

Rural Development, Regional Planning, Biogeography, Economic Geography, Urban Geography, Latin América. Post Doctoral Training: -

Four months in France for a Stage in Urbanism and Land Management, 1983

Current Positions:

Full Professor of Geography, Pontificia Universidad Católica del Perú (PUCP), since 1992. - Retired Professor of Geography at the Escuela Académica Profesional de Geografía, Universidad Nacional Mayor de San Marcos (UNMSM) since 2004. - Executive Director of the Center for Research in Applied Geography (CIGA-PUCP), since 1985. - Member of the Board, Sociedad Geográfica de Lima, since 1990. - Member of the Academic Board of Revista del Departamento de Geografía, National University of Tucumán, Faculty of Philosophy and Letters, since 2006. - Member of the Academic Board of Geotrópico, electronic journal published in, Colombia, since 2009. - Member of the Task Force Team on Megacities International Geographical Union (IGU) since 2000. - Editor of Boletín de la Sociedad Geográfica de Lima, since 2011. - Coordinator of the Cooperation Agreement between the Municipality of Frías, Piura and PUCP for the sustainable rural development since 2011.

Scholarships:

Latin American Scholarship Program of American Universities (LASPAU), 1970-1972. - Organization of American States (OAS) 1976-1977 and 1981-1982. - Centre International des Etudiants et Stagiaires (CIES), France, 1983 (4 months). - Fulbright Scholar-in-Residence at the University of Vermont, U.S.A., 1990, (6 months). - Visiting Professor at the Department of Geography, University of Bergen, Noruega, 1998 (3 months). - Guggenheim Fellowship, 1999.

Honors:

"Enlaces Award" given by the Conference of Latin Americanist Geographers (CLAG) in January, 2000.

INDEX